SpringerBriefs in Physics

T0223389

For further volumes:
http://www.springer.com/series/8902

Roger Boudet

Quantum Mechanics in the Geometry of Space–Time

Elementary Theory

 Springer

Roger Boudet
Honorary Professor
Université de Provence
Av. de Servian 7
34290 Bassan
France
e-mail: boudet@cmi.univ-mrs.fr

ISSN 2191-5423   e-ISSN 2191-5431

ISBN 978-3-642-19198-5   e-ISBN 978-3-642-19199-2

DOI 10.1007/978-3-642-19199-2

Springer Heidelberg Dordrecht London New York

Cover design: eStudio Calamar, Berlin/Figueres

Printed on acid-free paper

Springer is part of Springer Science+Business Media (www.springer.com)

Preface

The aim of the work we propose is a contribution to the expression of the present particles theories in terms entirely relevant to the elements of the geometry of the Minkowski space–time $M = \mathcal{R}^{1,3}$, that is those of the Grassmann algebra $\wedge \mathcal{R}^4$, scalars, vectors, bivectors, pseudo-vectors, pseudo-scalars of $\mathcal{R}^4$ associated with the signature $(1, 3)$ which defines $\mathcal{M}$, and, at the same time, the elimination of the complex language of the Pauli and Dirac matrices and spinors which is used in quantum mechanics.

The reasons for this change of language lie, in the first place, in the fact that this real language is the same as the one in which the results of experiments are written, which are necessarily real.

But there is another reason certainly more important. Experiments are generally achieved in a laboratory frame which is a galilean frame, and the fundamental laws of Nature are in fact independent of all galilean frame. So the theories must be expressed in an invariant form. Then geometrical objects appear, whose properties give in particular a clear interpretation of what we call energy. Also gauges are geometrically interpreted as rings of rotations of sub-spaces of local orthonormal moving frames. The energy–momentum tensors correspond to the product of a suitable physical constant by the infinitesimal rotation of these sub-spaces into themselves.

The passage of the expression of a theory from its form in a galilean frame to the one independent of all galilean frame, is difficult to obtain with the use of complex matrices and spinors language. The Dirac spinor which expresses the wave function Ψ associated with a particle is nothing else by itself but a column of four complex numbers. The definition of its properties requires actions on this column of the Dirac complex matrices.

An immense step in clarity was achieved by the real form ψ given in 1967 by David Hestenes (Oersted Medal 2002) to the Dirac Ψ. In this form, the Lorentz rotation which allows the direct passage to the invariant entities appears explicitly. In particular the geometrical meaning of the gauges defined by the complex Lie rings $U(1)$ and $SU(2)$ becomes evident.

It should be emphasized, like an indisputable confirmation of the independent work of Hestenes, that a geometrical interpretation of the Dirac Ψ had been implicitly given, probably during the years 1930, by Arnold Sommerfeld in a calculation related to hydrogenic atoms, and more generally and explicitly by Georges Lochak in 1956. In these works Ψ is expressed by means of Dirac matrices, these last ones being implicitly identified with the vectors of the galilean frame in which the Dirac equation of the electron is written.

But the use of a tool, the Clifford algebra $Cl(1,3)$ associated with the space $M = \mathcal{R}^{1,3}$, introduced by D. Hestenes, brings considerable simplifications. Pages of calculations giving tensorial equations deduced from the complex language may be replaced by few lines. Furthermore ambiguities associated with the use of the imaginary number $i = \sqrt{-1}$ are eliminated. The striking point lies in the fact that the "number i" which lies in the Dirac theory of the electron is a bivector of the Minkowski space–time M, a real object, which allows to define, after the above Lorentz rotation and the multiplication by $\hbar c/2$, the proper angular momentum, or spin, of the electron.

In the same aim, to avoid the ambiguousness of the complex Quantum Field Theory, due to the unseasonable association $i\hbar$ of $\hbar$ and i in the expression of the electromagnetic potentials "in quite analogy with the ordinary quantum theory" (in fact the Dirac theory of the electron), we give a presentation of quantum electrodynamics entirely real. It is only based on the use of the Grassmann algebra of M and the inner product in M.

The more the theories of the particles become complicated, the more the links which can unify these theories in an identical vision of the laws of Nature have to be made explicit. When these laws are placed in the frame of the Minkowski space–time, the complete translation of these theories in the geometry of space–time appears as a necessity. Such is the reason for the writing of the present volume.

However, if this book contains a critique, sometimes severe, of the language based on the use of the complex matrices, spinors and Lie rings, this critique does not concern in any way the authors of works obtained by means of this language, which remain the foundations of Quantum Mechanics. The more this language is abstract with respect to the reality of the laws of Nature, the more these works appear to be admirable.

Bassan, February 2011   Roger Boudet

Contents

**Part II   The U(1) Gauge in Complex and Real Languages.
Geometrical Properties and Relation with the Spin and
the Energy of a Particle of Spin 1/2**

**Part III   Geometrical Properties of the Dirac Theory
of the Electron**

Part VI   The Glashow–Salam–Weinberg Electroweak Theory

Part VII   On a Change of SU(3) into Three SU(2) × U(1)

Part IX Appendices

Chapter 1
Introduction

Abstract This chapter is devoted to the replacement of the complex matrices and spinors language by the use of the real Clifford algebra associated with the Minkowski space-time.

Keywords U(1) · SU(2) · Rotation groups · Clifford algebra

The following text is a step by step translation from the complex to a real language.

It contains the indication of all that this translation can bring to what is hidden in the first one, and of what may be unified in what appears as disparate in the presentation of the U(1), SU(2), SU(3) gauge theories (the third one being to be replaced by the direct product of three $SU(2) \times U(1)$). This translation leads to a better comprehension of what is called energy.

We have here made this contribution to the two theories widely verified by experiments, the electron and electroweak theories. Furthermore we propose an extension to the quarks chromodynamics theory (at present not entirely confirmed), with the condition of the possibility of a translation into the real language.

All that follows has been found simply by the use of the real Algebra of Space-Time (STA) [1], that is the Clifford algebra $Cl(M)$ associated with the Minkowski space M.

The use of this algebra was introduced for the theory of the electron in the fundamental article of David Hestenes [2] which introduces a real form for the Dirac spinor (the foundation of all the present theories of the particles).

Nevertheless the correspondence between the real and the complex languages will be recalled in detail.

Note that Hestenes has extended the use of the Clifford algebras to domains of physics other than quantum mechanics (see [3, 4]).

The following properties, established for these theories, are first based on the definition of an orthonormal moving frame $\{v, n_1, n_2, n_3\}$, defined at each point

R. Boudet, *Quantum Mechanics in the Geometry of Space–Time*,
SpringerBriefs in Physics, DOI: 10.1007/978-3-642-19199-2_1,
© Roger Boudet 2011

x of M, such that the time like vector v is colinear to a probability current $j = \rho v \in M(\rho > 0)$ of a particle.

The gauges $U(1)$ and $SU(2)$ are the groups of the rotations upon themselves of the plane generated by $\{n_1, n_2\}$, or "spin plane" [2], and the three-space $E^3(j)$ orthogonal to j generated by $\{n_1, n_2, n_3\}$ respectively.

A momentum-energy tensor $T = \rho T_0$ associated with one of these two gauges, is defined by a linear application $n \in M \to T_0(n) \in M$ which implies the product of a suitable physical constant by the *infinitesimal* rotation upon itself of the three-spaces generated by $\{v, n_1, n_2\}$ or $\{n_1, n_2, n_3\}$ respectively.

The trace of this energy-momentum tensor appears in the first term L_I of the lagrangian of the chosen theory.

Let us denote $a \cdot b$ (written $a^\mu b_\mu$ when a galilean frame $\{e_\mu\}$, $\mu = 0, 1, 2, 3$, is used) the scalar product of two vectors $a, b \in M$.

The second term L_{II} contains terms in the form $A \cdot j$, $B \cdot j$, and also sums $\sum W_k \cdot j_k$ ($j_k = \rho n_k$, $k = 1, 2, 3$) where A is an electromagnetic potential, B, $W_k \in M$ are vectors of space-time (bosons). Note that a term in the form $B \cdot j_i$ where j_i is isotropic appears in the electroweak theory.

The invariance in a gauge transformation implies a change in the expression of A and the bosons, related to the rotation of the spin plane or the three-space $E^3(j)$, and, in this last case, a change in the field associated with the bosons. These changes are well known. What is less or not at all known is the fact that these changes are related to the rotation of the spin plane and the three-space $E^3(j)$ in the case of $U(1)$ and $SU(2)$.

It does not seem possible to treat the $SU(3)$ gauge, as it is used in chromodynamics, with a complete interpretation in the geometry of M. But it is possible, without changing the standard lagrangian of the this theory, to replace this gauge by the direct product of three $SU(2) \times U(1)$ gauges, simply by the change of the eighth boson G_8 into $G_8/\sqrt{3}$, associated with the multiplication by $\sqrt{3}$ of the eighth Gell-Mann matrice followed by a suitable decomposition of this matrice into the sum of two matrices each one similar to the third isospin matrice.

If the chromodymanics theory, as it is, at present constructed, were confirmed by experiments, and if the possibility of the replacement of G_8 by $G_8/\sqrt{3}$ were infirm, one would face the following paradox. Two theories (Dirac electron, Glashow-Weinberg-Salam electroweak), widely confirmed by experiment, would be entirely enclosed in the geometry of space-time and a third one would not be enclosed.

On the contrary if this replacement were experimentally confirmed, the real space-time algebra would appear not only as a tool allowing noticeable simplifications in the calculations of confirmed theories (for the hydrogenic atoms see for example [5]), but also as a way to put in evidence fundamental properties of these theories, and at least like the indispensable language of quantum mechanics.

In an Addendum, we have given a geometrical construction of electromagnetism, which may be applied as well to the electromagnetic properties of charges endowed with a trajectory, like the ones of the charged particles in quantum mechanics whose the presence is based on a probability.

The aim of this Addendum is to show that the use of the complex Quantum Field Theory (QFT) is not necessary, and that the laws of the electromagnetism may be deduced from geometrical principles of an extreme simplicity.

Special attention has been brought to the construction of the Lorentz integral of the retarded potential, because some formulas which may be deduced from this integral play an important role in the theory of the hydrogen-like atoms [5].

However, concerning the interaction of the electron with a monochromatic electromagnetic wave in the photoeffect, we have followed what is generally used [6], but, we repeat, without the recourse to the QFT, that we replace, in a strict equivalent way, by a Real Quantum Electrodynamics. In addition to the simplicity, the reason of this replacement lies in the fact that the hidden use of unacceptable artifices, due to the unseasonable association $i\hbar$ of $\hbar$ and i in the expression of the potential, are suppressed.

References

1. D. Hestenes, *Space-Time Algebra*. (Gordon and Breach, New-York, 1966)
2. D. Hestenes, J. Math. Phys. **8**, 798 (1967)
3. D. Hestenes, Am. J. Phys. **71**, 104 (2003)
4. D. Hestenes, Ann. J. Math. Phys. **71**, 718 (2003)
5. R. Boudet, *Relativistic Transitions in the Hydrogenic Atoms*. (Springer, Berlin, 2009)
6. H. Bethe, E. Salpeter, *Quantum Mechanics for One or Two-Electrons Atoms*. (Springer, Berlin, 1957)

Part I
The Real Geometrical Algebra
or Space–Time Algebra. Comparison
with the Language of the Complex
Matrices and Spinors

Chapter 2
The Clifford Algebra Associated with the Minkowski Space–Time M

Abstract This Chapter is devoted to the first elements of the Clifford algebra $Cl(M)$ of the Minkowski space–time M when they are applied to quantum mechanics. A complete description of the Clifford algebra associated with all euclidean space lies in Chap. 14.

Keywords Grassmann algebra · Inner · Clifford products

2.1 The Clifford Algebra Associated with an Euclidean Space

The physicists construct their experiments in a particular galilean frame $\{e_\mu\}$, the laboratory frame. The objects and the equations expressing a theory are written in this frame.

However the laws of Nature are independent of this frame and entities associated with the particles are to be defined independently of all galilean frame. What is important is the Lorentz rotation which allows one the writing of these laws.

The recourse to matrices, which are generally used is much more complicated than the employment of the two following algebras. Furthermore some elements of the Grassmann algebra of M are relative to real objects which have an important physical meaning, as for example the proper angular momentum, or bivector spin of the electron.

In the language of the complex spinors, the imaginary number $i = \sqrt{-1}$, which lies in the Dirac equation of the electron, is nothing else but a bivector (*a real object!*) which defines, after the above Lorentz rotation and the multiplication by $\hbar c/2$, this angular momentum.

The first step in the use of these objects is the writing of the vectors independently of their components on a frame. Compare the writing $a^\mu b^\nu - b^\mu a^\nu$, called also "anti-symetric tensor of rank two", or simple bivector, with $a \wedge b$.

R. Boudet, *Quantum Mechanics in the Geometry of Space–Time*,
SpringerBriefs in Physics, DOI: 10.1007/978-3-642-19199-2_2,
© Roger Boudet 2011

The Clifford algebras are not very well known. But the Clifford algebra $Cl(E)$ of an euclidean space $E = \mathcal{R}^{p,n-p}$ is certainly the simplest algebra allowing the study of the properties of the orthogonal group $O(E)$ of E (see Sect. 14.5).

1. The definition and properties of the Grassmann algebra $\wedge \mathcal{R}^n$, of the p-vectors, elements of $\wedge^p \mathcal{R}^n$, and of the inner product between a p-vector and a q-vector in an euclidean space $E = \mathcal{R}^{p,n-p}$, are recalled in detail in Chap. 14.

   We simply mention here that a p-vector of E is nothing else but what the physicists call "an antisymmetric tensor of rank p" which is expressed by means of the components in a frame of E of the vectors of E which define this p-vector. But the use of the p-vectors does not need the recourse to a frame as it is shown below.

   We will denote by $A_p \cdot a$, $a \cdot A_p$ the inner products of a p-vector A_p by a vector a of E which correspond to the operation so-called (by the physicists) "contraction on the indices". The product $a \cdot b (a, b \in E)$ defines the signature $R^{p,n-p}$ of E. We will use in particular the relation

$$(a \wedge b) \cdot c = (b \cdot c)a - (a \cdot c)b, \quad a, b, c \in E \tag{2.1}$$

   which defines a vector orthogonal to c, situated in the plane (a, b). It will be employed for the definition of a bivector of rotation. The relation

$$(B \cdot c) \cdot d = B \cdot (c \wedge d) \in R, \quad c, d \in E, \quad B \in \wedge^2 E \tag{2.2}$$

   will be also used.

2. The Clifford algebra $Cl(E)$ associated with an euclidean space E is a real associative algebra, generated by $\mathcal{R}$ and the vectors of E, whose elements may be identified to the ones of the Grassmann algebra $\wedge E$. Furthermore this algebra implies the use of the inner products in E.

   The Clifford product of two elements A, B of $Cl(E)$ is denoted AB and verifies the fundamental relation

$$a^2 = a \cdot a \in \mathcal{R}, \quad \forall a \in E \tag{2.3}$$

   We simply mention in this chapter the properties we need, the complements lie in Chap. 14.

   If p vectors $a_i \in E$ are orthogonal their Clifford product verifies

$$a_1...a_p = a_1 \wedge ... \wedge a_p, \quad (a_k \in E, \ a_i \cdot a_j = 0 \text{ if } i \neq j) \tag{2.4}$$

In particular

$$a_1, a_2 \in E, \ a_1 \cdot a_2 = 0 \Rightarrow a_1 \cdot a_2 = a_1 a_2 = a_1 \wedge a_2 = -a_2 \wedge a_1 = -a_2 a_1 \tag{2.5}$$

The even sub-algebra $Cl^+(E)$ of $Cl(E)$ is composed of the sums of scalars and elements $a_1...a_p$ such that p is even.

One can immediately deduce from (2.4) that, using an orthonormal frame of E, the corresponding frame of $Cl(E)$ may be identified with the frame of $\wedge E$ and that $\dim(Cl(E)) = \dim(\wedge E) = 2^n$, $\dim(Cl^+(E)) = 2^{n-1}$.

One uses the following operation called "principal anti automorphism", or also "reversion",

$$A \in Cl(E) \to \tilde{A} \in Cl(E) \text{ so that } (AB)\tilde{} = \tilde{B}\tilde{A}$$
$$\tilde{\lambda} = \lambda, \ \tilde{a} = a, \ \lambda \in \mathcal{R}, \ a \in E \tag{2.6}$$

2.2 The Clifford Algebras and the "Imaginary Number" $\sqrt{-1}$

Let $\{e_1, e_2\}$ be a positive orthonormal frame of $\mathcal{R}^{2,0}$. We can write

$$(e_2 \wedge e_1)^2 = (e_2 e_1)^2 = (-e_1 e_2) e_2 e_1 = -(e_1)^2 (e_2)^2 = -1 \tag{2.7}$$

So a square root of -1 may be interpreted like a bivector of $\mathcal{R}^{2,0}$, a real object!

$Cl^+(2, 0)$ may be identified with the field $\mathcal{C}$ of the complex numbers. Note that the real geometrical interpretation of $\mathcal{C}$, in particular as allowing the definition of the rotations in the plane (e_1, e_2) had been found much more before the invention of the Clifford algebras. For example, using (2.1) one has

$$(e_2 \wedge e_1) \cdot e_1 = e_2, \qquad (e_2 \wedge e_1) \cdot e_2 = -e_1$$

which corresponds to the rotation of the vector of the frame through an angle of $\pi/2$. In $Cl(M)$ this relation may be written

$$i' e_1 = e_2 e_1^2 = e_2, \ i' e_2 = (-e_1 e_2) e_2 = -e_1 e_2^2 = -e_1, \ i' = e_2 e_1$$

Let $\{e_\mu\}$ a positive orthonormal frame, or galilean frame, of M be. We can write also

$$(e_2 \wedge e_1)^2 = (e_2 e_1)^2 = -1 \tag{2.8}$$

The "number" i which appears in the Dirac equation of a spin-up electron (for the spin-down i is replaced by $-i$) in its writing relative to this frame is nothing else but the bivector $e_2 \wedge e_1$ (see [1, 2]).

This bivector is, after the Lorentz rotation which makes this equation independent of all galilean frame and a multiplication by $\hbar c/2$, the bivector spin $(\hbar c/2)(n_2 \wedge n_1)$ of the electron that is, multiplied by physical constants, a pure real geometrical object.

Using the same method one can easily establish that another geometrical object [1], in which the square in $Cl(M)$ is also equal to -1 plays an important but quite different role. This object is the following 4-vector (in fact independent of all orthonormal frame, fixed or moving, of M)

$$\underline{i} = e_0 \wedge e_1 \wedge e_2 \wedge e_3 = e_0 e_1 e_2 e_3 \in \wedge^4 M, \text{ so that } \underline{i}^2 = -1 \qquad (2.9)$$

It corresponds in physics to the i of the writing $F = \mathbf{E} + i\mathbf{H}$ of the bivector electromagnetic field $F \in \wedge^2 M$.

So two quite different real geometrical objects, playing a fundamental role in the particles theories, are represented in the complex language by the same "imaginary number" i!

Let us denote

$$\mathbf{e}_k = e_k \wedge e_0 = e_k e_0, \quad k = 1, 2, 3 \qquad (2.10)$$

Applying (2.2), (2.1) one deduces

$$\mathbf{e}_k \cdot \mathbf{e}_j = (e_k \wedge e_0) \cdot (e_j \wedge e_0) = -e_k \cdot e_j$$
$$-e_k \cdot e_j = 0 \text{ if } k \neq j, \quad -e_k \cdot e_k = 1, \ k, j = 1, 2, 3$$

and so these bivectors of M may be considered as a frame of a space $E^3(e_0) = \mathcal{R}^{3,0}$, and also, as it easy to establish by using $\mathbf{e}_k = e_k e_0$,

$$\underline{i} = e_0 e_1 e_2 e_3 = \mathbf{e}_1 \mathbf{e}_2 \mathbf{e}_3 = \mathbf{e}_1 \wedge \mathbf{e}_2 \wedge \mathbf{e}_3 \in \wedge^3 E^3(e_0), \ \underline{i}^2 = -1 \qquad (2.11)$$

Since the $\mathbf{e}_k$ and the $\underline{i}\mathbf{e}_k$ may be considered as bivectors of $\mathcal{R}^{1,3}$ one deduces that $Cl^+(1, 3)$ may be identified with the ring of the Clifford biquaternions $Cl(3, 0)$.

The writing $F = \mathbf{E} + \underline{i}\mathbf{H}$ in $E^3(e_0)$ of $F \in \wedge^2 M$ corresponds to the definitions (2.10), (2.11).

2.3 The Field of the Hamilton Quaternions and the Ring of the Biquaternion as $Cl^+(3, 0)$ and $Cl(3, 0) \simeq Cl^+(1, 3)$

Hamilton introduced in its theory of the quaternions three objects i, j, k whose square is equal to -1, so that a quaternion q is in the form

$$q = d + ia + jb + kc, \quad a, b, c, d \in \mathcal{R}, \ i^2 = j^2 = k^2 = -1 \qquad (2.12)$$

verifying

$$i = -jk, \quad j = -ki, \quad k = -ij \qquad (2.13)$$

It was the first example of the fact that different objects are such that their square is equal to -1.

In fact i, j, k may be written in the form (with a change of sign with respect to the initial presentation by Hamilton)

$$i = \mathbf{e}_2 \wedge \mathbf{e}_3 = \mathbf{e}_2 \mathbf{e}_3 = \underline{i}\mathbf{e}_1, \quad j = \mathbf{e}_3 \wedge \mathbf{e}_1 = \mathbf{e}_3 \mathbf{e}_1 = \underline{i}\mathbf{e}_2$$

$$k = \mathbf{e}_1 \wedge \mathbf{e}_2 = \mathbf{e}_1\mathbf{e}_2 = \underline{i}\mathbf{e}_3 \tag{2.14}$$

and their squares in $Cl(3, 0)$ is equal to -1, in such a way that one can write $q \in Cl^+(3, 0)$.

Furthermore (2.13) may be deduced from this interpretation. For example

$$i = -jk = -(\mathbf{e}_3\mathbf{e}_1)(\mathbf{e}_1\mathbf{e}_2) = \mathbf{e}_2\mathbf{e}_3$$

One can write

$$q = d + \underline{i}\mathbf{a} \in Cl^+(3, 0) \text{ with } d \in \mathcal{R}, \ \underline{i}\mathbf{a} \in \wedge^2 E^3(e_0) \tag{2.15}$$

The biquaternions may be written

$$Q = q_1 + \underline{i}q_2 \in Cl(3, 0), \quad q_1, q_2 \in Cl^+(3, 0) \tag{2.16}$$

also as a consequence of (2.10), (2.11)

$$Q \in Cl^+(1, 3) \quad \rightarrow \quad Cl(3, 0) \simeq Cl^+(1, 3) \tag{2.17}$$

The $Cl^+(3, 0)$ of the Hamilton quaternions, *the only field* which may be associated with an euclidean space for $n > 2$, a privileged algebraic object, which gives a privileged place to the space of signature (3,0). This field plays an important role in the theory of the hydrogenic atoms (see [3, p. 930, 4, 5]).

Completing a sentence of the philosopher Kant one can say "The three-space in which we live is a certitude *algebraically* apodictic".

Considering the ring $Cl(3, 0)$ like the algebraic continuation of the field $Cl^+(3, 0)$ and thus $Cl^+(1, 3) \Leftrightarrow Cl(3, 0)$ like the algebraic continuation on the space $\mathcal{R}^{1,3}$ of the field $Cl^+(3, 0)$ of the Hamilton quaternions (see [6]), one can deduce that the signature $(1, 3)$ of the Minkowski space–time is privileged too. It is a very agreeable coincidence between pure data of the human mind and the laws of Nature. Furthermore, there is another important gift of Nature: the Dirac wave function, which is used not only in the electron theory but also in all the theories of the elementary particles, quarks and leptons, is, when it is written in real language, an element of this privileged ring $Cl^+(1, 3) \Leftrightarrow Cl(3, 0)$.

References

1. D. Hestenes, J. Math. Phys. **8**, 798 (1967)
2. D. Hestenes, J. Math. Phys. **14**, 893 (1973)
3. A. Sommerfeld, *Atombau und Spectrallinien*. (Fried. Vieweg, Braunschweig, 1960)
4. R. Boudet, C.R. Ac. Sc. (Paris) **278** A, 1063 (1974)
5. R. Boudet, *Relativistic Transitions in the Hydrogenic Atoms*. (Springer, Berlin, 2009)
6. D. Hestenes, *Space–Time Algebra*. (Gordon and Breach, New-York, 1966)

Chapter 3
Comparison Between the Real and the Complex Language

Abstract This Chapter is devoted to the definition of the Hestenes spinor and its comparison with the Dirac spinor.

Keywords Lorentz rotation · Takabayasi angle · Moving frame

3.1 The Space–Time Algebra and the Wave Function Associated with a Particle: The Hestenes Spinor

Hestenes has demonstrated in [1] that the wave function, considered in a galilean frame $\{e_\mu\}$, associated with an electron could be expressed as a biquaternion element ψ of $Cl^+(M)$ and written in the form ([1], Eq. 4.4) (see App. B)

$$\psi = \sqrt{\rho}e^{\underline{i}\beta/2}R \qquad (3.1)$$

$$\rho > 0, \quad \beta \in \mathcal{R}, \quad R\tilde{R} = \tilde{R}R = 1, \tilde{R} = R^{-1}, \quad \underline{i} \in \wedge^4 M, \quad \underline{i}^2 = -1$$

In fact one can write

$$\psi\tilde{\psi} \in Cl^+(M) \Rightarrow \psi\tilde{\psi} = \lambda + B + \underline{i}\mu, \quad \lambda, \mu \in R, \quad B \in \wedge^2 M$$

and from $(\psi\tilde{\psi})\tilde{} = \psi\tilde{\psi}, \tilde{B} = -B, \tilde{\underline{i}} = \underline{i}$, we deduce $B = 0$ and

$$\psi\tilde{\psi} = \lambda + \underline{i}\mu = \rho e^{\underline{i}\beta}, \quad \frac{\psi\tilde{\psi}}{\rho e^{\underline{i}\beta}} = 1, \quad R = \frac{\psi}{\sqrt{\rho}e^{\underline{i}\beta/2}} \Rightarrow R\tilde{R} = \tilde{R}R = 1$$

So R verifies $\tilde{R} = R^{-1}$ and corresponds to a representation of $SO^+(M)$ in $Cl^+(M)$, that is a Lorentz rotation.

The definition of $\underline{i}$ has been given in Eq. 2.9. The scalar ρ expresses the invariant probability density. The "angle" β does not intervene in what follows (its role is evoked below) and may be eliminated in the construction of the following currents.

R. Boudet, *Quantum Mechanics in the Geometry of Space–Time*,
SpringerBriefs in Physics, DOI: 10.1007/978-3-642-19199-2_3,
© Roger Boudet 2011

Since $a\underline{i} = -\underline{i}a$ if $a \in M$, $a \exp(\underline{i}\beta/2) = (\exp(-\underline{i}\beta/2)a$, and

$$a \in M \Rightarrow b = Ra\tilde{R} \in M, \quad \psi a\tilde{\psi} = \rho b \in M$$

one can write

$$\psi e_0 \tilde{\psi} = \rho v = j, \quad v^2 = 1, \quad \rho > 0 \tag{3.2}$$

The time-like vector j is the probability current.

Three other currents may be defined

$$\psi e_k \tilde{\psi} = j_k = \rho n_k, \quad n_k^2 = -1, \quad k = 1, 2, 3 \tag{3.3}$$

The vectors n_k play an important role in the theory of the electron, and, associated with three bosons, in the electroweak theory, also in the presentation that we propose of the chromodynamics theory.

Note that, in replacement of the Dirac spinor, an expression similar to Eq. 3.1 has been written in [2] by the employment of the Dirac matrices, but carries more complications in the use of ψ.

We shall call the biquaternion ψ a Hestenes spinor when it is written in the form of Eq. 3.1 *and applied to the study of quantum mechanics.*

Given all the applications in Physics of $Cl^+(1, 3)$, Hestenes has given to this ring the name of Space–Time Algebra (STA) [3].

In the gauge theories the density ρ does not interverne. Only the vectors n_1, n_2 in the $U(1)$ gauge, the three vectors n_k in the $SU(2)$ one. They play also a role in the definition of the momentum–energy tensors (see Sects. 5.2 and 8.2.2).

Note on the "angle" β. The role of the Yvon–Takabayasi–Hestenes "angle" β, which concerns not the vectors but the bivectors of M is obscure.

This "angle" plays no role in the gauge theories but gives an interpretation of the link electron–positron more satisfactory than the use of the T transform of the CPT invariance in the passage from the equation of the electron to the one of its associated positron.

We show in Sect. 7.2 that this interpretation proposed by Takabayasi (see [4], Eq. 10.3_b) is in fact a necessity.

The scalar β was introduced in the theory of the electron by Yvon [5], used by Lochak [2] in its expression of the wave function of the electron, studied in detail by Takabayasi [4], and independently rediscovered by Hestenes [1].

The presence in a physical theory of this scalar β is "strange" (as said Louis de Brogie) with respect to ρ (probability) and R (Lorentz rotation) widely used in the standard presentation of quantum mechanics. But, because it is an indisputable component of a biquaternion, element of $Cl^+(1, 3)$, when this biquaternion is considered as containing a Lorentz rotation R, its necessity is confirmed by the place played by the biquaternions in physics. We recall that the biquaternions were introduced by Sommerfeld (see [6]) in the study of the hydrogenic atoms.

We have shown in [7] that β has a value non null, though small, in the solution of the hydrogen atom, except in the plane $x^3 = 0$.

One can give to this entity a geometrical interpretation by considering $G = \exp(i\beta/2)R$ as defining a group G of transformations $X \to GX\tilde{G}$ (see [8]) that we have called the Hestenes group.

Because $\tilde{i} = i$, $a\underline{i} = -\underline{i}a$, $a \in M$, this group is reduced to the group of the Lorentz rotations when X is a vector but it defines a "rotation" if X is a bivector and the transformation $\beta \to \beta + \pi$ allows one to inverse the orientation of a simple bivector.

If this "angle" plays a role in the passage from a particule to the definition of its antiparticule, it has no incidence on the gauge theories, for the reason that they are relative to the rotations of sub-frames of M and so imply only sub-groups of $SO^+(1, 3)$.

3.2 The Takabayasi–Hestenes Moving Frame

The role of the vector j, and so v [given by Eq. 3.2 in STA] in the theory of a particle whose wave function is a Dirac spinor, is well known.

On the other hand the role of the vectors n_k, except the fact that the bivector spin of the electron is in the form $(\hbar c/2)n_2 \wedge n_1$ ("up") or $(\hbar c/2)n_1 \wedge n_2$ ("down"), is ignored in the standard study of the particles theories. An exception : it appears in the works on the electron by the Louis de Broglie school during the 1950s (see in particular [9]) in which was introduced the local orthonormal frame

$$F = \{v, n_1, n_2, n_3\}, \quad v = Re_0\tilde{R}, \quad n_k = Re_k\tilde{R} \tag{3.4}$$

called by Habwachs [9] the Takabayasi moving frame and considered independently by Hestenes in [1].

These vectors n_k also play a fundamental role in the definition of invariant entities, the gauge theories and the invariant definition of the energy–momentum tensors.

The ring of the finite rotations upon themselves of the sub-frame of F, (n_1, n_2) and (n_1, n_2, n_3) are directly related to the $U(1)$ and $SU(2)$ gauges, their infinitesimal rotations to the definition of momentum–energy tensors. It is impossible to understand the geometrical nature of what we call energy in particles theories without the consideration of these sub-frames (see Chap. 16).

3.3 Equivalences Between the Hestenes and the Dirac Spinors

In addition to $\gamma_\mu \Leftrightarrow e_\mu$, justified in Chap. 15 by Eq. 15.9, one can deduce the equivalences, not at all evident (see Sect. 15.4), established for the first time by Hestenes [1],

$$\Phi \Leftrightarrow Q \to \gamma^\mu \Phi \Leftrightarrow e^\mu Q e_0 \tag{3.5}$$

where Φ may be considered as a Dirac spinor, as far as it is a column of four complex numbers upon which the Dirac matrices act, and Q is a biquaternion element of $Cl^+(1, 3)$.

Furthermore, in the theories of the spin 1/2 particles one uses the equivalence

$$i\Psi = \Psi i \Leftrightarrow \psi e_2 e_1 = \psi \underline{i} e_3 e_0 = \psi \underline{i} e_3, \quad i = \sqrt{-1} \Leftrightarrow e_2 e_1 = \underline{i} e_3 \qquad (3.6)$$

where ψ is a Hestenes spinor, which is in addition to Eq. 3.5 the key of the translation in STA of the Dirac spinor in the theory of the electron.

We recall that the change of i into $e_2 e_1$ corresponds to the change of i into $\gamma_2 \gamma_1$ already used by Sommerfeld [6] and Lochak [2] in an form given to the Dirac spinor in which the γ_μ correspond implicitly to the e_μ.

In the theories implying the $SU(2)$ gauge we have replaced the equivalence Eq. 3.6 by

$$i\Psi = \Psi i \Leftrightarrow \underline{i}\psi = \psi \underline{i}, \quad i = \sqrt{-1} \Leftrightarrow \underline{i} \qquad (3.7)$$

The Dirac current $j \in M$ associated with a Dirac spinor Ψ is given by the equivalence (see Sect. 15.4)

$$j^\mu = \bar{\Psi}\gamma^\mu\Psi \in \mathcal{R} \Leftrightarrow j = j^\mu e_\mu = \psi e_0 \tilde{\psi} \in M \qquad (3.8)$$

3.4 Comparison Between the Dirac and the Hestenes Spinors

Something is crucial in quantum mechanics : the passage of a theory, expressed with respect to a galilean frame, to a form invariant with respect to all galilean frame. Only such a form can give the meaning of the terms of the theory directly with respect to the space–time.

So the Lorentz rotation R which allows this passage plays a fundamental role.

As we have evoked above, the Dirac spinor Ψ is nothing else but a column of four complex numbers and cannot contain by itself a Lorentz rotation. The presence of such a rotation and its use require the employment of the Dirac matrices. They allow one to determine the current of the probability density j, also the invariant density and then the unit vector v.

But the calculation of the other unit vectors is much more difficult with the use of the Dirac spinors.

The determination of the vectors n_k is obtained by one line in STA, that is Eq. 3.3, followed by the division by $\rho = \sqrt{j^2}$.

References

1. D. Hestenes, J. Math. Phys. **8**, 798 (1967)
2. G. Jakobi, G. Lochak, C.R. Ac. Sc. (Paris) **243**, 234 (1956)
3. D. Hestenes, *Space–Time Algebra* (Gordon and Breach, New-York, 1966)

4. T. Takabayasi, Supp. Prog. Theor. Phys., **4**, 1 (1957)
5. J. Yvon, J. Phys. et le Radium **VIII**, 18 (1940)
6. A. Sommerfeld *"Atombau und spectrallinien"* (Fried. Vieweg, Braunschweig, 1960)
7. R. Boudet, C. R. Acad. Sci. (Paris) **278**, (1974)
8. R. Boudet, in *"Clifford Algebras and their Applications in Mathematical Physics"*, A. Micali, R. Boudet, J. Helmstetter, (eds.) (Kluwer, Dordrecht, 1992) p. 343
9. F. Halbwachs, Théorie relativiste de fluides à spin. (Gauthier-Villars, Paris, 1960)

Part II
The U(1) Gauge in the Complex and Real Languages. Geometrical Properties and Relation with the Spin and the Energy of a Particle of Spin 1/2

Chapter 4
Geometrical Properties of the U(1) Gauge

Abstract A change of gauge and the condition of a gauge invariance, as well on the wave function of a particle, as upon a potential vector acting on the particle, is recalled for the complex language and established for the real one.

Keywords U(1) · SO$^+$ (1, 3) · Finite · Infinitesimal rotations · Energy

4.1 The Definition of the Gauge and the Invariance of a Change of Gauge in the U(1) Gauge

4.1.1 The U(1) Gauge in Complex Language

Let us denote by Ψ the Dirac spinor (defined by a column of four complex numbers upon which the Dirac matrices can act) associated with a particle. A change of gauge is defined by the transform

$$\Psi \rightarrow \Psi' = U\Psi = \Psi U, \quad U = e^{i\chi/2}, \quad i = \sqrt{-1}, \quad \chi \in \mathcal{R} \qquad (4.1)$$

where, we recall

$$i\Psi = \Psi i \qquad (4.2)$$

The number χ may be fixed or dependent on the point x of M and in this cases the change of gauge is to be said global or local.

4.1.2 The U(1) Gauge Invariance in Complex Language

Let us consider, associated to a particle submitted to a potential $A \in M$, an expression in the form

R. Boudet, *Quantum Mechanics in the Geometry of Space–Time*,
SpringerBriefs in Physics, DOI: 10.1007/978-3-642-19199-2_4,
© Roger Boudet 2011

$$L_\mu = \partial_\mu \Psi i - g A_\mu \Psi, \tag{4.3}$$

where Ψ is a Dirac spinor expressing the wave function of the particle and g a suitable physical constant.

In the Dirac theory of the electron one has $g = q/(\hbar c)$ where $q = -e, e > 0$, is the charge of the electron and so $g A_\mu$ has the dimension of the inverse of a length.

In a change of a local gauge like Eq. 4.1, L_μ becomes

$$L'_\mu = \left[\partial_\mu \Psi i - \left(\frac{1}{2} \partial_\mu \chi + g A_\mu \right) \Psi \right] U \tag{4.4}$$

If L'_μ is chosen in such a way that $L'_\mu = L_\mu U$, nothing is changed, except the transform of Ψ into ΨU, if A_μ is changed into

$$A'_\mu = A_\mu - \frac{\partial_\mu \chi}{2g}, \quad A' = A - \frac{\partial \chi}{2g} \in M, \quad \partial = e^\mu \partial_\mu \in M \tag{4.5}$$

Such a change of the potential A does not affect the field $F = \partial \wedge A$ associated with A, because $\partial \chi$ is a gradient.

4.1.3 A Paradox of the U(1) Gauge in Complex Language

The definition Eq. 4.1 of the U(1) gauge in complex language leads to a paradox. Since i is considered as nothing else but the number $\sqrt{-1}$, a change of gauge must be interpreted as related to some abstract property of the wave function Ψ, and, as a consequence, the number χ has no geometrical meaning. But in a change of gauge it must be considered with a geometrical meaning, because it appears in addition to a potential which is a vector of the Minkowski space–time M, and so as a gradient in M, that is a geometrical object.

Thus a geometrical interpretation of the gauge $U(1)$ appears as a necessity.

4.2 The U(1) Gauge in Real Language

A geometrical interpretation of the $U(1)$ gauge as a sub-group of $SO^+(1, 3)$ has been implicitly contained in [1], explicitly and independently described in [2, 3]. But it is unknown to most physicists and sometimes violently negated by some of them. Their reason is related to the meaning of the number i in the theory of the electron and their ignorance of the relation (4.6): since $i\Psi = \Psi i$ they say that it is *impossible* that $U(1)$ may be interpreted as corresponding to a sub-group of $SO^+(1, 3)$.

4.2.1 The Definition of the U(1) Gauge in Real Language

The Dirac wave function being expressed in the form Eq. 3.1 given by Hestenes [3], Eq. 4.4 transforms Eq. 4.1 into

$$\psi \rightarrow \psi' = \psi U, \quad U = e^{e_2 e_1 \chi/2}, \quad \chi \in \mathcal{R} \tag{4.6}$$

where

$$e^{e_2 e_1 \chi/2} = \cos(\chi/2) + \sin(\chi/2) e_2 e_1$$

Note that ψ is to be multiplied *on the right* by $U = e^{e_2 e_1 \chi/2}$.

But here, what is called a change of gauge U(1) in complex language, corresponds in STA to

$$U = e^{e_2 e_1 \chi/2}, \quad R \rightarrow R' = RU = Re^{e_2 e_1 \chi/2} \tag{4.7}$$

which induces a rotation through an angle χ in the plane (n_2, n_1):

$$n_2' = \cos\chi\, n_2 + \sin\chi\, n_1, \quad n_1' = -\sin\chi\, n_2 + \cos\chi\, n_1 \tag{4.8(1)}$$

with

$$R' e_0 R'^{-1} = e_0, \quad R' e_3 R'^{-1} = e_3 \tag{4.8(2)}$$

So the replacing of $i\chi/2$ by $e_2 e_1 \chi/2$ gives to χ the real geometrical meaning of an angle.

Furthermore, since the plane (n_2, n_1) (in "up", or (n_1, n_2) in "down") is the plane defined by the bivector spin of a particle of spin 1/2, one can give to the $U(1)$ gauge the following definition:

The $U(1)$ gauge is the ring of the rotations upon itself of the plane defined by the bivector spin of a particle of spin 1/2.

4.2.2 The U(1) Gauge Invariance in Real Language

The replacing of $i = \sqrt{-1}$ by the bivector $e_2 \wedge e_1 = e_2 e_1$ in Eq. 4.3 gives

$$L_\mu = \partial_\mu \psi e_2 e_1 - g A_\mu \psi, \tag{4.9}$$

But here $e_2 e_1$ is necessarily to be written on the right of ψ, when one can write $\Psi i = i\Psi$. Apart from this difference, the calculation is the same as in Sect. 4.1.2 and gives also Eq. 4.5 but here the paradox evoked above disappears.

A change of gauge through an angle χ implies

$$\omega \to \omega' = \omega - \partial \chi \in M \tag{4.10}$$

where

$$\omega_\mu = \partial_\mu n_2.n_1 = -\partial_\mu n_1.n_2 \tag{4.11}$$

which expresses the infinitesimal rotation upon itself of the plane (n_2, n_1).

The invariance is achieved by the change of the potential A

$$A \to A' = A - \frac{\partial \chi}{2g} \tag{4.12}$$

as in Eq. 4.5 but with a meaning of χ given by Eqs. 4.8, 4.10, 4.11, whose meanings are purely geometrical.

But because $\partial \chi$ is implied in a change of the potential, we have a hint of the fact that the vector ω is, multiplied by a suitable physical constant ($\hbar c/2$ in the case of the electron), related to the energy of the particle.

It is a first indication of the role played by the infinitesimal rotation upon itself of the plane (n_2, n_1) in the geometrical interpretation of the energy associated with the particle.

As established in Sects. 5.3 and 6.6 for the electron we will be able to insure (see [4]) that

The infinitesimal rotation upon itself of the plane defined by the bivector spin of a particle of spin 1/2 defines, multiplied by a suitable physical constant, ($\hbar c/2$ in the case of the electron), the energy of the particle.

References

1. G.G. Jakobi, G. Lochak, C. R. Acad. Sci. (Paris) **243**, 234 (1956)
2. F. Halbwachs, Théorie relativiste de fluides á spin (Gauthier-Villars, Paris, 1960)
3. D. Hestenes, J. Math. Phys. **8**, 798 (1967)
4. R. Boudet, C. R. Acad. Sci. (Paris) **272A**(767), (1971)

Chapter 5
Relation Between the U(1) Gauge, the Spin and the Energy of a Particle of Spin 1/2

Abstract Real language allows one to put in evidence the relation between the U(1) Gauge, the Spin and the Energy of a particle of spin 1/2.

Keywords Spin plane · Infinitesimal rotation · Momentum-energy

5.1 Relation Between the U(1) Gauge and the Bivector Spin

In the theory of the electron, the plane (n_1, n_2) has been called by Hestenes [1], the "spin plane" because the bivector spin, or proper angular momentum, of the electron is in the form "spin up",

$$\sigma = \frac{\hbar c}{2} n_2 \wedge n_1 \tag{5.1}$$

In "spin down" $n_2 \wedge n_1$ is replaced by $n_1 \wedge n_2$.

The relations between the $U(1)$ gauge and the bivector spin of the electron, clearly established in [2] and discovered independently in [1] is absent in the standard use of the complex formalism.

5.2 Relation Between the U(1) Gauge and the Momentum–Energy Tensor Associated with the Particle

The momentum–energy tensor associated with a particle in the $U(1)$ gauge implies, for its values v, n_1, n_2, a linear application from M in M in the form (see Chap. 16)

$$n \in M \rightarrow N(n) = (\Omega_\mu \cdot (\underline{i}(s \wedge n))) e^\mu \tag{5.2}$$

R. Boudet, *Quantum Mechanics in the Geometry of Space–Time*,
SpringerBriefs in Physics, DOI: 10.1007/978-3-642-19199-2_5,
© Roger Boudet 2011

where

$$\Omega_\mu = 2(\partial_\mu R)R^{-1}, \ s = n_3 = Re_3R^{-1}$$

which is to be multiplied by ρg_1, where g_1 is a suitable physical constant (equal to $\hbar c/2$ in the case of the electron).

Thus N verifies (16.10)

$$N(v) = (\partial_\mu n_2 \cdot n_1)e^\mu = \omega \tag{5.3}$$

$$N(n_1) = (\partial_\mu n_2 \cdot v)e^\mu, \ \ N(n_2) = -(\partial_\mu n_1 \cdot v)e^\mu \tag{5.4}$$

which expresses the infinitesimal rotation of the sub-frame $\{v, n_2, n_1\}$ upon itself but in such a way that a change of gauge only affects the infinitesimal rotation of the plane (n_2, n_1).

5.3 Relation Between the U(1) Gauge and the Energy of the Particle

The energy E of the electron in a galilean frame $\{e_\mu\}$ is defined by the projection on e_0 of the vector $(\hbar c/2)\omega$ (see [3, 4])

$$E = \frac{\hbar c}{2}N(v) \cdot e_0 = \frac{\hbar c}{2}\omega_0 \tag{5.5}$$

So one obtains a remarkable geometrical interpretation of the energy of the electron: it is the product of physical constants by the infinitesimal rotation of the "spin plane" upon itself.

That gives a hint of the fact that in general a gauge is closely related to the infinitesimal rotation of a local sub-frame upon itself.

The relation between the $U(1)$ gauge and the interpretation of the energy is absent in the use of the complex formalism.

References

1. D. Hestenes, J. Math. Phys. **8**, 798 (1967)
2. F. Halbwachs, J.M. Souriau, J.P. Vigier, J. Phys. et le radium **22**, 393 (1967)
3. R. Boudet, C.R. Ac. Sc. (Paris) **272** A, 767 (1971)
4. R. Boudet, C.R. Ac. Sc. (Paris) **278** A, 1063 (1974)

Part III
Geometrical Properties of the Dirac Theory of the Electron

Chapter 6
The Dirac Theory of the Electron in Real Language

Abstract This section is devoted to: (a) the real form given by Hestenes to the Dirac equation, (b) the properties of the Dirac theory which may be deduced when this equation is written with respect to a particular galilean frame, (c) the principal invariant entities associated with this equation.

Keywords Current · Bivector spin · Tetrode tensor · Lagrangian

6.1 The Hestenes Real form of the Dirac Equation

In order to avoid all ambiguity concerning the charge of the electron (see [1], p. 98) in the presentation of the Dirac equation, we will denote by $q = -e$ the charge of the electron, with $e > 0$ (as for example in [2], p. 77 and Eq. 5.17).

One can pass immediatly from the Dirac equation in the galilean frame $\{e_\mu\}$

$$\hbar c \gamma^\mu \partial_\mu (i\Psi) - mc^2 \Psi - q A_\mu \gamma^\mu \Psi = 0, \quad i = \sqrt{-1}, \quad q = -e, \; (e > 0) \quad (6.1)$$

where $\partial_\mu = \partial/\partial x^\mu$, to the form given to this equation in ([3], Eq. 2.15),

$$\hbar c e^\mu \partial_\mu \psi e_2 e_1 e_0 - mc^2 \psi - q A \psi e_0 = 0, \quad A = A_\mu e^\mu \in M \quad (6.2)$$

by using Eqs. 3.6, then 3.5, with $Q = \partial_\mu \psi e_2 e_1$. Note that $e_2 e_1 e_0$ may be written $\underline{i} e_3$.

Multiplying on the right by e_0 we obtain a form more appropriated (see Sect. 7.1)

$$\hbar c e^\mu \partial_\mu \psi e_2 e_1 - mc^2 \psi e_0 - q A \psi = 0, \quad A = A_\mu e^\mu \in M \quad (6.3)$$

Note that each term of this equation has the dimension of an energy.

R. Boudet, *Quantum Mechanics in the Geometry of Space–Time*,
SpringerBriefs in Physics, DOI: 10.1007/978-3-642-19199-2_6,
© Roger Boudet 2011

One can find e_1e_2 in place of e_2e_1 in this equation ([4], Eq. 5.1). These two possibilities correspond to the states "up" (e_2e_1) and "down" (e_1e_2) of the electron and are related with the orientation of the bivector spin.

We will work only with Eq. 6.3, the second equation giving results which may be easily deduced from the change of e_2e_1 into e_1e_2.

6.2 The Probability Current

The probability current is given by Eq. 3.2. It is used with the following normalization

$$\psi e_0 \tilde{\psi} = \rho v = j, \quad \rho > 0, \quad v^2 = 1, \quad \int j^0(\mathbf{r})d\tau = 1, \quad j^0 = j \cdot e^0 \qquad (6.4)$$

where the integration of j^0 is made in all the three-space $E^3(e_0)$.

6.3 Conservation of the Probability Current

The probability current j verifies the relation

$$\partial \cdot j = 0, \quad \partial = e^\mu \partial_\mu \qquad (6.5)$$

In fact one has

$$\partial \cdot j = [e^\mu \partial_\mu(\psi e_0 \tilde{\psi})]_S = [e^\mu(\partial_\mu \psi)e_0 \tilde{\psi}]_S + [\psi e_0(\partial_\mu \tilde{\psi})e^\mu]_S = I + \tilde{I}$$

where $[X]_S$ means the scalar part of X, and the relation $[e^\mu Y]_S = [Ye^\mu]_S$ has been applied.

So one can deduce from Eq. 6.2 after multiplication on the right by $e_1e_2\tilde{\psi}$

$$I = [e^\mu(\partial_\mu \psi)e_0 \tilde{\psi}]_S = \frac{1}{\hbar c}[mc^2 \psi e_1 e_2 \tilde{\psi} + qA\psi e_0 e_1 e_2 \tilde{\psi}]_S$$

We can write

$$I = \frac{\rho}{\hbar c}[mc^2 e^{i\beta} n_1 n_2 + qi An_3]_S$$

where $vn_1n_2 = -\underline{i}n_3$, $A\underline{i} = -\underline{i}A$ has been used. So I is null, as is its reversion $\tilde{I}$, because no term of the right hand of this equation is a scalar.

6.4 The Proper (Bivector Spin) and the Total Angular–Momenta

The bivector spin

$$\sigma = \frac{\hbar c}{2} n_2 \wedge n_1 \tag{6.6}$$

is easily calculated as explained in Sect. 3.3 .

It defines a plane (n_2, n_1) which was called, we recall, by Hestenes [4] the "spin plane".

The total angular–momentum is defined as

$$J = x \wedge p + \sigma \tag{6.7}$$

where p is defined by Eq. 6.11.

6.5 The Tetrode Energy–Momentum Tensor

The energy–momentum (Tetrode) tensor T [5] of the Dirac electron is a linear application of M into M, written by Hestenes [3] in STA

$$n \in M \to T(n) = \hbar c \left(\left[\partial_\mu \psi \underline{i} e_3 \tilde{\psi} n \right]_S e^\mu \right) - \rho(v \cdot n) q A \in M \tag{6.8}$$

where $\underline{i} e_3 = e_0 e_2 e_1$.

The Hestenes form of T is justified by the correspondence Eq. 6.13 below concerning the trace of this tensor which lies in the lagrangian of the Dirac equation.

We have shown in [6] that

$$\left[\partial_\mu \psi \underline{i} e_3 \tilde{\psi} n \right]_S e^\mu = \frac{\rho}{2} (N(n) - (n.s) \partial \beta), \quad \underline{i} e_3 = e_0 e_2 e_1 \tag{6.9}$$

where (Eq. C.9)

$$N(n) = (\Omega_\mu \cdot (\underline{i}(s \wedge n))) e^\mu, \quad \Omega_\mu = 2(\partial_\mu R) \tilde{R}, \quad s = n_3 = R e_3 \tilde{R}$$

In fact, a simple calculation shows that we can write

$$\partial_\mu \psi \underline{i} e_3 \tilde{\psi} n = \frac{1}{2} (\rho(\Omega_\mu \underline{i} - \partial_\mu \beta) + \underline{i} \partial_\mu \rho) sn$$

from which, since in particular

$$[\Omega_\mu \underline{i} sn]_S = \Omega_\mu \cdot (\underline{i}(s \wedge n)), \quad [\underline{i}(\partial_\mu \rho) sn]_S = 0$$

we deduce Eq. 6.9.

Writing

$$T(n) = \rho(T_0(n) - (n \cdot v)qA), \quad T_0(n) = \frac{\hbar c}{2}(N(n) - (n \cdot s)\partial\beta) \qquad (6.10)$$

we have, applying Eq. 5.2, the so-called energy–momentum vector

$$p = T_0(v) - qA = \frac{\hbar c}{2}\omega - qA, \quad \omega = (\partial_\mu n_2.n_1)e^\mu \qquad (6.11)$$

which is, as shown in Sect. 4.2.2, gauge invariant.

A form similar to Eq. 6.10 of the Tetrode tensor, including the presence of $\partial\beta$, has been explicited by Halbwachs [7], but with a mechanical interpretation of $N(n)$ different from our geometrical one.

The trace of $\rho T_0(n)$ is

$$\rho T_0(e^\nu) \cdot e_\nu = \hbar c \left[\partial_\mu \psi e_0 e_2 e_1 \tilde{\psi} e^\nu\right]_S \delta^\mu_\nu = \hbar c \left[\partial_\mu \psi e_0 e_2 e_1 \tilde{\psi}\right]_V \cdot e_\mu \qquad (6.12)$$

where $[X]_V$ means the vector part of $X \in Cl(M)$.

So the trace of the tensor ρT_0 appears in the lagrangian of the Dirac electron following the correspondence

$$\bar{\Psi}\gamma^\mu i\partial_\mu \Psi \iff \left[e^\mu(\partial_\mu \psi e_2 e_1)e_0\tilde{\psi}\right]_S = e^\mu \cdot \left[\partial_\mu \psi e_0 e_2 e_1 \tilde{\psi}\right]_V \qquad (6.13)$$

deduced from the two correspondences Eqs. 3.6 then 3.5.

The presence of $e_0 e_2 e_1$ in the lagragian is a hint on the fact that the momenum–energy tensor contains the expression of an infintesimal rotation of the sub-frame (n_0, n_2, n_1).

6.6 Relation Between the Energy of the Electron and the Infinitesimal Rotation of the "Spin Plane"

As a confirmation of what we said in Sect. 5.3, the vector

$$\frac{\hbar c}{2}\omega = p + qA, \quad \omega = (\partial_\mu n_2 \cdot n_1)e^\mu$$

is such that ω_0 is the energy E of the electron in the galilean frame $\{e_\mu\}$:

$$E = \frac{\hbar c}{2}\omega_0, \quad \omega_0 = \partial_0 n_2.n_1 \qquad (6.14)$$

We have by a direct calculation verified in [6] this property for the hydrogen atom.

6.7 The Tetrode Theorem

The Tetrode theorem [5] is the following:

"The space–time divergence of the energy–momentum tensor of the Dirac electron is equal to the density of the Lorentz force acting on the electron".

Let us replace the vectors n by vectors e^ν of the frame $\{e_\mu\}$. One can write

$$\partial_\nu T^\nu = \rho(q F.v), \quad T^\nu = T(e^\nu), \quad F = e^\nu \wedge \partial_\nu A \tag{6.15}$$

Chapter. 18 contains a STA proof of the Tetrode theorem which allows one to shorten the proof given by Tetrode [5], published in 1928 (just after the article of Dirac !).

6.8 The Lagrangian of the Dirac Electron

Multiplying Eq. 6.2 on the right by $\tilde{\psi}$ and taking the scalar part one has, because $[\psi\tilde{\psi}]_S = \cos\beta$, $\rho v = \psi e_0 \tilde{\psi}$

$$L = \hbar c e^\mu \cdot [(\partial_\mu \psi) e_0 e_2 e_1 \tilde{\psi}]_V - mc^2 \cos\beta - A \cdot (q\rho v) = 0 \tag{6.16}$$

which is the lagrangian of the Dirac electron and is strictly equivalent to that of the conventional formalism. It is null when the Dirac equation is satisfied.

6.9 Units

The only constants we will use are the three fundamental constants (revised in 1989 by B. N. Taylor):

(1) the speed of light $c = 2.99792458 \times 10^{10}$ cm sec^{-1}.
(2) the electron charge magnitude $e = 4.803\ 206 \times 10^{-10}$ (e.s.u.)
(3) the reduced Planck constant $\hbar = h/2\pi = 1.054\ 572 \times 10^{-27}$ erg sec. In addition we will use
(4) the electron mass $m = 9.109\ 389 \times 10^{-28}$ g. All the other constants used will be derived from these four ones, in particular
(5) the fine structure constant

$$\alpha = \frac{e^2}{\hbar c} = \frac{1}{137.035\ 989} \quad (e \text{ in e.s.u.}) \tag{6.17}$$

and as unit of length:
(6) the "radius of first Bohr orbit"

$$a = \hbar^2/(me^2) = \hbar/(mc\alpha) = 5.291\ 772 \times 10^{-9} \text{ cm} \tag{6.18}$$

Note. In other respects one introduces in the expression of the electromagnetic potentials the factor $1/(4\pi\epsilon_0)$ (the presence of 4π is due to the writing $4\pi j^\mu$ instead of j^μ in the current term of the Maxwell equations) where ϵ_0 is the permitivity of free space, and e is expressed in e.m.u:

$$\epsilon_0 = 8.854187 \times 10^{-12} \, \text{F m}^{-1}, \quad e = 1.602 \, 1777 \times 10^{-19} \, (\text{e.m.u.})$$

That gives (with c expressed in metres) the same value of α with the expression

$$\alpha = \frac{e^2}{4\pi\epsilon_0\hbar c}, \quad (e \text{ in e.m.u.}) \tag{6.19}$$

For simplicity and to be in agreement with the largest part of the reference articles and treatises mentioned here, we will use the former expressions of the potentials and the constant α, in preference to these letter ones.

References

1. L.D. Landau, E.M. Lifshiftz, Quantum Mechanichs, vol. 4. ((Pergamon Press), New York, 1971)
2. F. Halzen, D. Martin, *Quarks and Leptons.* (Wiley, USA, 1984)
3. D. Hestenes, J. Math. Phys. **14**, 893 (1973)
4. D. Hestenes, J. Math. Phys. **8**, 798 (1967)
5. H. Tetrode, Z. f. Phys. **49**, 858 (1928)
6. R. Boudet, C R Ac Sc (Paris) **278**A, 1063 (1974)
7. F. Halbwachs, *Théorie relativiste de fluides á spin.* (Gauthier-Villars, Paris, 1960)

Chapter 7
The Invariant Form of the Dirac Equation and Invariant Properties of the Dirac Theory

Abstract This section is relative to the invariant form of the Dirac equation and some fundamental invariant properties of the Dirac theory which may be deduced from this form.

Keywords Invariance · Positron · Broglie · Lorentz · Einstein formulas

7.1 The Invariant Form of the Dirac Equation

Multiplying on the right Eq. 6.3 first by $e_2 e_1$ then by ψ^{-1}:

$$\hbar c e^{\mu} \partial_{\mu} \psi \psi^{-1} = -(mc^2 \psi e_0 + q A \psi) e_2 e_1 \psi^{-1} \tag{7.1}$$

where $\psi^{-1} = R^{-1} \exp(-i\underline{\beta}/2)/\sqrt{\rho}$, we have the following invariant form of the Dirac equation [1]

$$\frac{\hbar c}{2}(e^{\mu}\Omega_{\mu} + \partial\beta\underline{i} + \partial(ln\rho)) = -(mc^2 e^{i\beta}v + q A)\sigma_0 \in \wedge^1 M \oplus \wedge^3 M \tag{7.2}$$

$$\Omega_{\mu} = 2(\partial_{\mu}R)R^{-1}, \quad \sigma_0 = n_2 n_1 = n_2 \wedge n_1$$

This equation corresponds to the state "spin up". For the state "spin down", σ_0 is to be changed into $-\sigma_0$.

We recall that each bivector Ω_{μ} represents the infinitesimal rotation of the "Takabayasi–Hestenes proper frame" $\{v, n_1, n_2, n_3\}$ when the point x moves in the e_{μ} direction.

This equation may be divided into two parts:

1. The $\wedge^3 M$ part D_I, four real equations implying seven real scalars R, β, is independent of ρ. These scalars, associated with the physical constants $\hbar, c, q$ and the potential A, lead to the construction of all the entities (energy, spin) which are observable. Note that D_I is gauge invariant.

R. Boudet, *Quantum Mechanics in the Geometry of Space–Time*,
SpringerBriefs in Physics, DOI: 10.1007/978-3-642-19199-2_7,
© Roger Boudet 2011

2. The vector part D_{II}, four equations implying eight real scalars R, β, ρ, implies in addition the density ρ which has a probabilistic (or, following the authors, statistical) meaning.

So one can deduce a particularity of the Dirac theory which may be extended to the other ones: these theories may be expressed by means of equations implying the observable physical entities, but these equations contain too many real parameters with respect to the numbers of equations, to be solved and other equations implying probabilistic (or statistical) parameters appear as a necessity.

About the link between equations D_I and D_{II} we have established in [2] the following theorem:

D_{II} is implied by D_I and the three conservation relations

$$\partial_\mu(\rho v^\mu) = 0, \quad \partial_\mu(T^{\mu\nu}) = \rho f^\nu, \quad \partial_\mu(S^{\mu\nu\xi}) = (T^{\xi\nu} - T^{\nu\xi})$$

where T is the Tetrode tensor, $f \in M$ the Lorentz force and $S = \rho v \wedge \sigma$.

In particular cases of the choice of the potential A, particular solutions of the equation D_I, may lead to the expression of phenomena directly observable (see Sects. 7.4, 7.5).

7.2 The Passage from the Equation of the Electron to the One of the Positron

The conditions of the invariance of the Dirac equation when one considers the positron associated with an electron whose orientation of the spin is given (here in Eq. 7.1 the one of the bivector $n_2 n_1 = n_2 \wedge n_1$) are well known with the standard operations on the usual presentation of the Dirac equation. These conditions are obtained by the CPT transforms.

It is easy to obtain the CPT invariance by using Eq. 7.2.

(a) C (Charge) changes q in $-q = e > 0$.
(b) P (Parity) changes (e_2, e_1) into (e_1, e_2) and so $n_2 n_1$ into $n_1 n_2$.
(c) T (Time reversion) changes e_0 into $-e_0$ and so v in $-v$.

The left hand part of Eq. 7.2 is unchanged by (a), (b) and (c):

– As a consequence of (b), σ_0 is changed into $-\sigma_0$, so $-q(-\sigma_0) = q\sigma_0$ and the charge term in Eq. 7.2 is unchanged.
– As a consequence of (c), $-v(-\sigma_0) = v\sigma_0$, and the mass term in Eq. 7.2 is unchanged.

So the right hand part of Eq. 7.2 is unchanged. The left hand part is unchanged by any of these transforms.

But the T transformation seems imply that the positrons come from the future.

In order to explaining this particularity of the T transformation, where the positrons could be considered as coming from the future, Stückelberg (1941), then

Feynman (1948) proposed an interpretation of the T transformation based on "the idea that a negative which propagates backward in time, or equivalently a positive energy antiparticle propagating forward in time." (see [3], p. 77). Such supposition is based on the relation of v with the charge $-e < 0$ in the Dirac equation of the electron and on the relation of $-v$ with the charge $e > 0$ in the associated equation of the positron.

In [4], Eq. 10.3$_b$, Takabayasi avoids the change ot v into $-v$ by the following transform:

(c)' *The angle β is changed into $\beta + \pi, v$ remaining unchanged.*

As an additional justification of the Takabayasi transformation, one can remark that:

1. The "angle" β concerns the "rotation" of bivectors, not of vectors, and its change into $\beta + \pi$, implying the reversal of a bivector, is coherent with the change $n_2 n_1 = n_2 \wedge n_1$ into $n_1 n_2 = n_1 \wedge n_2$ implied by the P transformation, which associates to an electron a positron whose the orientation of the bivector spin is opposite.
2. The angle β appears in the mass term of the lagragian of the positron in the form $-mc^2 \cos \beta$ where $0 \leq \beta \leq \pi$, and it is possible that the transformation of β into $\beta + \pi$, which is in fact the change m for the electron into $-m$, that is a negative mass for the positron, is related with the disappearance of the mass in the process of an annihilation electron–positron.

7.3 The Free Dirac Electron, the Frequency and the Clock of L. de Broglie

The equation of the free Dirac electron may be deduced from the equation D_I simply by supposing that $A = 0$ and furthermore that $\beta = 0$.

Multiplying Eq. 7.2 on the left by $\sigma_0 = n_2 n_1$, then taking the vector part of this new equation, we have

$$[e^\mu \Omega_\mu n_2 n_1]_V = e^\mu((\Omega_\mu . n_2) . n_1) = (\partial_\mu n_2 . n_1) e^\mu = \omega, \quad n_2 n_1 = n_2 \wedge n_1$$

and we arrive at

$$\frac{\hbar c}{2} \omega = mc^2 v, \quad \omega = (\partial_\mu n_2 . n_1) e^\mu \qquad (7.3)$$

Considering the galilean frame where the electron is at rest, we can write $v = e_0$, and furthermore $x^0 = ct$ gives the proper time t of the free electron.

So the energy of the free electron is

$$E = \frac{\hbar c}{2} \omega . v = mc^2, \quad v = e_0 \qquad (7.4)$$

The introduction of the L. de Broglie frequency

$$\nu_0 = \frac{mc^2}{h} = \frac{1}{2\pi} \cdot \frac{c(\omega.v)}{2}, \quad \omega = (\partial_\mu n_2.n_1)e^\mu \tag{7.5}$$

which is so related to the infintesimal rotation of the spin plane upon itself, allows us to give a geometrical picture of what is called the L. de B. clock.

Let us denote

$$n_1 = \cos\varphi\, e_1 + \sin\varphi\, e_2, \quad n_2 = -\sin\varphi\, e_1 + \cos\varphi\, e_2 \tag{7.6}$$

which corresponds to $R = \exp(-e_2 e_1 \varphi/2)$.

The angle φ is only function of x_0 and one deduces

$$\omega = e^0 \frac{d\varphi}{dx^0}, \quad c\omega = \frac{d\varphi}{dt} v$$

Equation 7.5 becomes

$$\nu_0 = \frac{1}{2}\frac{d\varphi}{dt} \cdot \frac{1}{2\pi} \tag{7.7}$$

So we can give to the hand of the L. de Broglie clock the following pure geometrical interpretation. It is a vector N of the spin plane such that

$$N = \cos(\varphi/2)\, e_1 + \sin(\varphi/2)\, e_2 \tag{7.8}$$

which runs on the direct direction of the plane (n_1, n_2), that is, on the dial of the clock, anti-clockwise.

In the case where the spin is "down", one can see in the same way that the vector N is in the form

$$N = \cos(\varphi/2)\, e_1 - \sin(\varphi/2)\, e_2 \tag{7.9}$$

which runs on the inverse direction of the plane (n_1, n_2), that is, in the dial of the clock, clockwise.

Passage to the equation of the positron. Because the charge q does not intervene (it is sufficient that $A = 0$, a fact which does not forbid to an electron to exist!) the C transform is not to be considered. The P transform changes in Eq. 7.3 ω into $-\omega$ and so its right-hand part into $-mc^2 v$. So it is necessary to suppose that v is changed in $-v$ and that the positron comes from the future, an hypothesis envisaged by Stückelberg and Feynman, but which cannot be taken into account because the charge is absent in the present case, in such a way that this possibility is invalidated, or that m is changed into $-m$ in agreement with the Takabayasi transform $\beta = 0 + \pi = \pi$.

7.4 The Dirac Electron, the Einstein Formula of the Photoeffect and the L. de Broglie Frequency

We consider the particular, but important case that we have considered in [5], where the potential A is in the form

$$A = A^0 e_0 + A^3 e_3 = 0, \quad A^0 = g(x^0, x^3), \quad A^3 = f(x^0, x^3) \tag{7.10}$$

We can have a solution of the equation D_I by assuming that $\beta = 0$ and that the spin plane keeps a fixed direction in such a way that $\sigma_0 = n_2 \wedge n_1$ is defined by

$$n_1 = \cos\phi \, e_1 + \sin\phi \, e_2, \quad n_2 = -\sin\phi \, e_1 + \cos\phi \, e_2$$

Since we have then $A.\sigma_0 = 0$, $A\sigma_0 = A \wedge \sigma_0$, multiplying as before Eq. 7.2 on the left by $\sigma_0 = n_2 n_1$, then taking the vector part of this new equation, we have the equation

$$\frac{\hbar c}{2}\omega - qA = mc^2 v, \quad \omega = (\partial_\mu n_2.n_1)e^\mu \tag{7.11}$$

similar to Eq. 7.3, but with a potential in addition, and also deduced here from the D_I equation.

The time component of this equation is

$$\frac{\hbar c}{2}\omega^0 - W = mc^2 v^0, \quad W = qA^0 \tag{7.12}$$

where

$$\frac{\hbar c}{2}\omega^0 = h\nu_1, \quad \nu_1 = \frac{1}{2}\frac{\partial\phi}{\partial t}\cdot\frac{1}{2\pi} \tag{7.13}$$

Now, assuming the approximation

$$mc^2 v^0 = mc^2\left[1 - \frac{\mathbf{v}^2}{c^2}\right]^{-1/2} \simeq mc^2 + \frac{1}{2}m\mathbf{v}^2 \tag{7.14}$$

and writing mc^2 in the form $h(mc^2/h) = h\nu_0$ one deduces

$$h\nu - W = \frac{1}{2}m\mathbf{v}^2, \quad \nu = \nu_1 - \nu_0 \tag{7.15}$$

that is the formula of the photoeffect introduced by Einstein in 1905.

An important point is to be noticed: the hidden presence of the L. de Broglie frequency mc^2/h in the Einstein formula which has been the starting point, after the discovery in 1900 by Planck of the quantum of energy $h\nu$, of the quantum theory of the electron.

7.5 The Equation of the Lorentz Force Deduced from the Dirac Theory of the Electron

Considering the same particular case as above, we follows a way with hypothesis similar to the ones used by Hestenes in [6] and [7], in particular in the fact that the angle β is null.

Taking the spacetime curl of Eq. 7.11, and *since ω_μ is a gradient*, one obtains

$$-qF = mc^2(\partial \wedge v), \quad F = \partial \wedge A) \tag{7.16}$$

We notice that $\hbar$ *is eliminated* during this operation and so we go towards a classical theory of the electron.

One can write

$$-qF.v = mc^2(\partial \wedge v).v \tag{7.17}$$

with, since $(\partial_\mu v).v = (\partial_\mu(v^2)/2 = 0$

$$(\partial \wedge v).v = (e^\mu \wedge \partial_\mu v).v = -(v.\partial)v$$

one has

$$m(V \cdot \partial)V = \frac{q}{c}F \cdot V, \quad V = cv \tag{7.18}$$

Now we make the point x as describing one of the current lines C, defined in the spacetime plane (e_0, e_3), by Eq. 7.11. One has along C

$$V = \frac{dx}{d\tau}, \quad (V.\partial)V = \frac{dV}{d\tau} = \frac{d^2x}{d\tau^2} \tag{7.19}$$

where τ is the proper time parameter of C, where $(V.\partial)V$ corresponds to the derivative of V along the vector V tangent at the point x to the curve C, that is the acceleration of the particle.

Equation 7.18 becomes (see [6], Eqs. 3.5)

$$m\frac{dV}{d\tau} = \frac{q}{c}F \cdot V, \quad V = \frac{dx}{d\tau} \tag{7.20}$$

that is an equation which has exactly the same form as the Lorentz force equation and, so, may be considered as defining "a trajectory".

However Eq. 7.20 is to be considered as compatible with Eq. 7.11 whose corresponding integral line defines a current line of the Dirac theory.

Moreover, if we consider C as a trajectory, the plane orthogonal at x to C (which keeps a fixed direction parallel to the (e_1, e_2) plane), is nothing else but the "spin plane" of the Dirac theory.

In this way we can say that the Planck constant, which appears in Eq. 7.11, and the spin are hidden parameters of the classical theory of the electron.

But one has to notice that the Dirac theory *is reduced here to its dynamical equation the equation* D_I. The part of the Dirac theory, equation D_{II} which implies the density ρ has not been taken into account.

Furthermore in all that precedes the spacetime curve C is to be considered in the particular case where it is situated in a spacetime plane and where the space curve of the electron is a straight line.

7.6 On the Passages of the Dirac Theory to the Classical Theory of the Electron

In the particular case of a potential in the form of Eq. 7.10, the two points which are implied in the classical theory of the electron, that is the absence of h and of the probability density ρ, have been deduced *exactly* from the Dirac equation, limited to the dynamic equation D_I.

In the case of other forms of the potential A, such that the direction of the spin plane varies sufficiently slowly in such a way that ω may be considered as a gradient, and so h is eliminated, one can deduce from D_I *as an approximation* the classical behaviour of the electron in presence of the field $F = \partial \wedge A$.

And now, we can associated what precedes with the sentence of Einstein, "God cannot play with dice", about the difference between a classical and a quantum electron. On one side a particle which may be clearly situated in the space, which has a trajectory, and on the other an object whose position seems to depend on hazard. In agreement with the conviction of Einstein, we can say that, in what precedes, nothing allows one to assert that hazard is to be associated with this object, which the Dirac height scalars equations cannot specify exactly its position. Four of these equations are relative to seven scalar variables corresponding to observable entities, and so the determination of these entities only by means of these equations is impossible. But and it is only a deficiency of our knowledge, not some caprice of Nature. One scalar variable and four equations more, and the calculation of these entities is possible, except the position which may be only approached in a statistical point of view.

However the case treated above of an electron both classical, because it has a trajectory, and quantal, because it satisfies the Einstein formula of the photoeffect, appears as a window open to the hitherto inaccessible reality.

References

1. R. Boudet, CR. Ac. Sc. (Paris). **272**(767), 272A, 767 (1971)
2. R. Boudet, J. Math. Phys. **26**, 718 (1985)
3. v. Halzen, D. Martin, *Quarks and leptons*. (J. Wiley and Sons, U.S.A., 1984)

4. T. Takabayasi, Supp. of the Prog. Theor. Phys. **4**, 1 (1957)
5. R. Boudet, in *"Bell's Theorem and the Foundations of Physics"*. ed. by A. van der Merwe, F. Selleri, G. Torozzi. (World Scientific, Singapore, 1992), p. 92
6. D. Hestenes, *Clifford Algebras and their Applications in Mathematical Physics*, ed. by J. Chisholm and A. Common (Reidel. Pub. Comp., Dordrecht, 1986) pp. 321–346
7. D. Hestenes, Am. J. Phys. **71**, 631 (2003)

Part IV
The SU(2) Gauge and the Yang–Mills Theory in Complex and Real Languages

Chapter 8
Geometrical Properties of the SU(2) Gauge and the Associated Momentum–Energy Tensor

Abstract A change of gauge and the condition of a gauge invariance, both on the wave function of a particle and upon a potential vector acting on the particle, is recalled for the complex language and established for the real one.

Keywords Isotriplets · Bivector · Three-space · Infinitesimal rotation

8.1 The SU(2) Gauge in the General Yang–Mills Field Theory in Complex Language

The Yang–Mills (Y.M.) lagrangian (see [1], p. 8)

$$L = L_I - gL_{II} : L_I = \bar{\Psi}\gamma^\mu i\partial_\mu \Psi, \quad L_{II} = \bar{\Psi}\gamma^\mu B_\mu \Psi \tag{8.1}$$

$$B_\mu = \frac{1}{2}W_\mu^k \tau_k = \frac{1}{2}\mathbf{W}_\mu$$

where τ_k are the isospin (or Pauli) matrices.

In this standard expression the set $\{\tau_k\}$ is interpreted as the frame $\{e_k\}$ of an "isotriplet space" isomorphic to $\mathcal{R}^{3,0}$. This space is to be considered as the space $E^3(e_0)$ generated by the bivectors of M

$$\mathbf{e}_k = e_k \wedge e_0 = e_k e_0$$

So the B_μ appear as bivectors of M and $W^k = e^\mu W_\mu^k$ as vectors of M.

A question is the nature of the two components of Ψ on which the matrices τ_k act. It is not specified in the treatises I know, except for the electroweak theory.

This question is discussed in [2], Sect. 4.1 that we recall in Chap. 17, and the conclusion is that in any case Ψ may be an "ordinary" Dirac spinor or Ψ a couple of Dirac spinors. The only case where the τ_k may be considered as matrices is the one where Ψ is a right or a left doublet in the $SU(2) \times U(1)$ gauge (see Chap. 9).

R. Boudet, *Quantum Mechanics in the Geometry of Space–Time*,
SpringerBriefs in Physics, DOI: 10.1007/978-3-642-19199-2_8,
© Roger Boudet 2011

So the SU(2) gauge can be considered as available alone only with the interpretation of the τ_k not like entities isomorphic to bivectors of M, but as really the geometrical bivectors defined by Eq. 8.6 below, with the use of the STA.

The conventional gauge $SU(2)$ transformation is achieved by the relation

$$B'_\mu = U B_\mu U^{-1} + \frac{i}{g}(\partial_\mu U)U^{-1}$$

where U belongs to a sub-group of $SU(2)$. Let us denote

$$\hat{\Omega}_\mu = 2(\partial_\mu U)U^{-1} \tag{8.2}$$

in such a way that

$$B'_\mu = \frac{1}{2}\mathbf{W}'_\mu, \quad \vec{W}'_\mu = U\mathbf{W}_\mu U^{-1} + \frac{i}{g}\hat{\Omega}_\mu \tag{8.3}$$

The associated Y.M. field is

$$F_{\nu\mu} = \partial_\nu B_\mu - \partial_\mu B_\nu - gi(B_\mu B_\nu - B_\nu B_\mu), \quad F'_{\nu\mu} = U F_{\nu\mu} U^{-1} \tag{8.4}$$

or, with the quality $\tau_k = \mathbf{e}_k$ of vectors of $E^3(e_0)$ (bivectors of M) given to the isospin matrices τ_k and applying the relation in $E^3(e_0)$

$$-\frac{i}{2}(\mathbf{ab} - \mathbf{ba}) = -\underline{i}(\mathbf{a} \wedge \mathbf{b}) = \mathbf{a} \times \mathbf{b}$$

where $\mathbf{a} \times \mathbf{b}$ means the vector product in E^3, one obtains

$$F_{\nu\mu} = \frac{1}{2}(\partial_\nu \mathbf{W}_\mu - \partial_\mu \mathbf{W}_\nu + g\mathbf{W}_\mu \times \mathbf{W}_\nu) \tag{8.5}$$

We remark on Eq. 8.3 that if $\mathbf{W}_\mu$ is a bivector, $\hat{\Omega}_\mu$ is a bivector too.

So the relation $F'_{\nu\mu} = U F_{\nu\mu} U^{-1}$ may be deduced from $\partial^2_{\nu\mu} U = \partial^2_{\mu\nu} U$ by means of the definition (8.2) which gives the relation Eq. C.19.

We have chosen to use $\hat{\Omega}_\mu = 2(\partial_\mu U)U^{-1}$ instead of $(\partial_\mu U)U^{-1}$ because it is already an indication of the geometrical meaning of the gauge.

8.2 The SU(2) Gauge and the Y.M. Theory in STA

8.2.1 The SU(2) Gauge and the Gauge Invariance in STA

All that follows could be simply replaced by supposing that U is considered like an element of $SO^+(M)$. But we prefer to treat, as in [2], Sect. 4.3, the question with the use of STA in detail because it brings important elements absent from the standard theory.

In the $SU(2)$ gauge, the role of the "isocurrents" j_k orthogonal to the probability current j of a particle, whose wave function is an invertible biquaternion ψ, is important.

We will use the concordances

$$\tau_k \Leftrightarrow \mathbf{e}_k = e_k \wedge e_0 = e_k e_0 \tag{8.6}$$

$$\bar{\Psi} \gamma^\mu \tau_k \Psi \Leftrightarrow e^\mu \cdot (\psi \mathbf{e}_k e_0 \tilde{\psi}) = e^\mu \cdot (\psi e_k \tilde{\psi}) = j_k^\mu \tag{8.7}$$

which will be justified by what follows.

We deduce

$$L_{II} = \frac{1}{2} \bar{\Psi} \gamma^\mu \mathbf{W}_\mu \Psi \Leftrightarrow \frac{1}{2} W_\mu^k e^\mu \cdot (\psi e_k \tilde{\psi}) = \frac{1}{2} W^k \cdot j_k \tag{8.8}$$

where $\mathbf{W}_\mu = W_\mu^k e_k e_0 \in \wedge^2 M$ and $W^k \in M$.

We consider ψ instead of Ψ, and we denote by U a rotation such that

$$U e_0 U^{-1} = e_0, \quad R \to R' = RU \Rightarrow R' e_0 R'^{-1} = v \tag{8.9}$$

giving

$$\Omega_\mu \to \Omega'_\mu = \Omega_\mu + R \hat{\Omega}_\mu R^{-1}, \quad R \mathbf{W}_\mu R^{-1} \to R(U \mathbf{W}_\mu U^{-1}) R^{-1} \tag{8.10}$$

The concordance with the complex L_1 requires an interpretation of $i \Psi$ (Eq. 3.7), that we have introduce in [3], different from the one of Eq. 3.6. Here we will write

$$i \Psi = \Psi i \Leftrightarrow \underline{i} \psi = \psi \underline{i}, \quad i = \sqrt{-1} \Leftrightarrow \underline{i} \tag{8.11}$$

Applying the concordances Eqs. (8.11), (3.5) we have

$$L_I = \bar{\Psi} \gamma^\mu i \partial_\mu \Psi \Longleftrightarrow \left[e^\mu \partial_\mu \psi \underline{i} e_0 \tilde{\psi} \right]_S = e^\mu \cdot \left[\partial_\mu \psi \underline{i} e_0 \tilde{\psi} \right]_V \tag{8.12}$$

We can write

$$\partial_\mu \psi \underline{i} e_0 \tilde{\psi} = \frac{1}{2} (\xi_\mu v + \rho \underline{i} \Omega_\mu v), \quad \xi_\mu = (\partial_\mu \rho + \underline{i} \rho \partial_\mu \beta) \underline{i} \tag{8.13}$$

where the ξ_μ are only function of ρ, β.

Note that $\psi \mathbf{W}_\mu e_0 \tilde{\psi} = \rho R \mathbf{W}_\mu R^{-1} R e_0 R^{-1} = \rho R \mathbf{W}_\mu R^{-1} v$,

$$L = e^\mu \cdot \left[\frac{1}{2} (\xi_\mu v + \rho J_\mu v) \right]_V, \quad J_\mu = \underline{i} \Omega_\mu - g R \mathbf{W}_\mu R^{-1} \tag{8.14}$$

The gauge transformation defined by U in the complex formalism is nothing else but a rotation upon itself of the three-space orthogonal at the point x of M to the probability current $j = \rho v$ of the Y.M. particle.

We obtain the geometrical interpretation of the gauge $SU(2)$:

The gauge $SU(2)$, associated with a particle, corresponds to the ring of the rotations in the three-space orthogonal to the time-like vector which is the probability current of the particle. It is a sub-group of $SO^+(1, 3)$.

The gauge transformation leaves v invariant but defines a rotation of the sub-frame upon $\{n_1, n_2, n_3\}$ upon itself, ρ, β, and so the ξ_μ, remaining unchanged.

We are going to find X such that the change

$$\mathbf{W}_\mu \to \mathbf{W}'_\mu = U\mathbf{W}_\mu U^{-1} + X$$

leaves J_μ, and so L, unchanged:

$$J_\mu = \underline{i}\Omega_\mu + R(\underline{i}\hat{\Omega}_\mu - g(U\mathbf{W}_\mu U^{-1} + X)R^{-1}) \tag{8.15}$$

which gives $X = (\underline{i}/g)\hat{\Omega}_\mu$ and

$$\mathbf{W}'_\mu = U\mathbf{W}_\mu U^{-1} + \frac{i}{g}\hat{\Omega}_\mu \tag{8.16}$$

exactly as in Eq. 8.3 of the complex formalism and with the same field $F_{\nu\mu}$.

What is new is the geometrical interpretation of a gauge transformation in SU(2): a rotation upon itself of the three space $E^3(v)$.

We see the geometrical link with the gauge $U(1)$ where the rotation is relative to the "spin plane".

8.2.2 A Momentum–Energy Tensor Associated with the Y.M. Theory

The part of a momentum–energy tensor associated with the Y.M. theory which does not take into account the B_μ may be written in STA

$$n \in M \to T(n) = g_1\left(\left[\partial_\mu \psi \underline{i} e_0 \tilde{\psi} n\right]_S e^\mu\right) \tag{8.17}$$

where g_1 is a suitable physical constant.

Since, in a $U(1)$ gauge, $\psi i e_3$ is replaced by $\psi i e_0$ in an $SU(2)$ gauge, we deduce (by application of the concordance (8.11) in place of (3.6)) a $SU(2)$ energy–momentum tensor by replacing, in a $U(1)$ energy–momentum tensor, $s = n_3$ by v. This tensor ρT_0 will be obtained by Eq. 6.10, with v in place of s, $S(N)$, Eq. 16.12, which expresses the infinitesimal rotation upon itself of $\{n_1, n_2, n_3\}$, replacing $N(n)$, and a suitable physical constant g_1 in place of $\hbar c$

$$n \in M \rightarrow \rho T_0(n) = \rho g_1 \left[\frac{1}{2}(S(n) - (n \cdot v)\partial\beta) \right] \in M. \qquad (8.18)$$

In the gauge $SU(2)$, the energy–momentum tensor contains the infinitesimal rotation of the three-space orthogonal to the probability current of the particle.

The trace of $T(e^\nu) \cdot e_\nu/g_1$ of $T(n)/g_1$ is

$$(e^\nu \cdot \left[\partial_\mu \psi \underline{i} e_0 \tilde{\psi} \right]_V) e^\mu \cdot e_\nu) = e^\nu \cdot \left[\partial_\mu \psi \underline{i} e_0 \tilde{\psi} \right]_V \delta_\nu^\mu = e^\mu \cdot \left[\partial_\mu \psi \underline{i} e_0 \tilde{\psi} \right]_V = L_I \quad (8.19)$$

in conformity with Eq. 8.12

With the introduction of the physical constant g_1 the Lagrangian will be written in the form

$$L' = g_1 L_I - g_2 L_{II}, \quad g = \frac{g_2}{g_1} \qquad (8.20)$$

8.2.3 The STA Form of the Y.M. Theory Lagrangian

We can deduce now from Eqs. 8.19, 8.8 the equivalence

$$L = L_I - gL_{II} \Leftrightarrow L = e^\mu \cdot \left[\partial_\mu \psi \underline{i} e_0 \tilde{\psi} \right]_V - \frac{g}{2} W^k \cdot j_k \qquad (8.21)$$

8.3 Conclusions About the $SU(2)$ Gauge and the Y.M. Theory

Considered separately the $SU(2)$ gauge and the Y.M. theory have no place in the complex language. The use of STA is a necessity.

The wave function Ψ, on which the isospin matrices act, cannot be either a Dirac spinor, couple of Pauli spinors, or a couple of Dirac spinors.

The isospin matrices may be interpreted like bivectors of M, and Ψ as corresponding to a Hestenes spinor ψ, invertible biquaternion.

They may be also associated with a bivector $\mathbf{a} + \underline{i}\mathbf{b}$ of M (see the three first lines of Table 1 of [4]).

In any case the $SU(2)$ gauge, considered as alone, cannot be considered as deduced from complex matrices like the isospin ones, and as associated with a particle of spin 1/2.

One can remark that, as they are presented in the "complex" language, Sect. 8.1, it is sufficient to consider the τ_k as bivectors of M and U not as belonging to a Lie group but to $SO^+(M)$.

The properties of the Y.M. theory remain with the use of the complex language in the part $SU(2)$ of $SU(2) \times U(1)$ lying in the electroweak theory and in the part implying the three first Gell-Mann matrices of the chromodynamics one.

In the first theory, the above ψ is a left doublet which may be an invertible biquaternion. It must be a doublet (left or right?) in the second one because of the spin 1/2 of the particles which are considered (see the Comments in Sect. 12.1).

If the $SU(2)$ gauge does not lie separately (to our knowledge) in the theories presently accepted, it is not impossible that it will be used for the explanation of some phenomena not yet known.

References

1. M. Carmeli, Kh. Huleihil, E. Leibowitz, *Gauge Fields* (World Scientific, Singapore, 1989)
2. R. Boudet, *Adv. Appl. Clifford Alg.* (Birkhaüser Verlag Basel, 2008), p. 43
3. R. Boudet, in *The Theory of the Electron*, ed. by J. Keller, Z. Oziewicz (UNAM, Mexico, 1997), p. 321
4. D. Hestenes, Space–time structure of weak and electromagnetic interactions. Found. Phys. **12**, 153–168 (1982)

Part V
The SU(2) × U(1) Gauge in Complex and Real Languages

Chapter 9
Geometrical Properties of the SU(2)× U(1) Gauge

Abstract After the definition of the left and right doublets associated with two wave functions, the role of the τ_k matrices acting on these entities and their geometrical interpretation is explained.

Keywords Doublets · SU(2) · U(1) · SU(2) × U(1)

9.1 Left and Right Parts of a Wave Function

In this chapter we recall the calculations achieved in [1].

We consider an "ordinary" spinor Ψ, that is a Dirac spinor which may be replaced by a Hestenes spinor ψ, and so a invertible biquaternion, and its decomposition

$$\Psi = \Psi_L + \Psi_R \tag{9.1}$$

$$\Psi_L = \frac{1}{2}(1 - \gamma^5)\Psi, \quad \Psi_R = \left(\frac{1}{2}(1 + \gamma^5)\right)\Psi \tag{9.2}$$

$$\gamma^5\Psi = \gamma^0\gamma^1\gamma^2\gamma^3 i\Psi, \quad i = \sqrt{-1} \tag{9.3}$$

(see [2], Eq. 5.49) in the so called left and right parts of Ψ.

The equivalences deduced from Eqs. 3.5, 3.6

$$\gamma^5 = \gamma^0\gamma^1\gamma^2\gamma^3 i\Psi \Leftrightarrow$$

$$e^0 e^1 e^2 e^3 \psi \underline{i} e_3 (e_0)^4 = -e_0 e_1 e_2 e_3 \psi \underline{i} e_3 = -\underline{i}\psi \underline{i} e_3 = \psi e_3$$

$$\frac{1}{2}(1 \mp \gamma^5)\Psi \Leftrightarrow \frac{1}{2}(1 \mp e_3)\psi \tag{9.4}$$

lead in STA to the decomposition

R. Boudet, *Quantum Mechanics in the Geometry of Space–Time*,
SpringerBriefs in Physics, DOI: 10.1007/978-3-642-19199-2_9,
© Roger Boudet 2011

$$\psi = \psi_L + \psi_R \tag{9.5}$$

$$\psi_L = \psi \frac{1}{2}(1 - e_3) = \psi u, \quad \psi_R = \psi \frac{1}{2}(1 + e_3) = \psi \tilde{u} \tag{9.6}$$

$$u = \frac{1}{2}(1 - e_3), \quad \tilde{u} = \frac{1}{2}(1 + e_3) \tag{9.7}$$

$$u + \tilde{u} = 1, \quad u\tilde{u} = \tilde{u}u = 0, \quad u^2 = u, \quad \tilde{u}^2 = \tilde{u} \tag{9.8}$$

Note that $u, \tilde{u}$ are idempotents.

Let us consider the two following spacetime vectors

$$j = \psi e_0 \tilde{\psi} = \rho v, \quad j' = \psi e_3 \tilde{\psi} = \rho s, \quad s = n_3 \tag{9.9}$$

We recall that the spacetime vectors j and j' are respectively the probability density and the "spin density" currents of the particle, whose wave function is ψ.

Note that $j \pm j' = \rho(v \pm s)$ are isotropic vectors: $(j \pm j')^2 = 0$.

Using

$$e_3 \tilde{u} = u e_3, \quad e_0 \tilde{u} = u e_0, \quad u e_3 = \frac{1}{2}(e_0 - e_3), \quad \tilde{u} e_0 = \frac{1}{2}(e_3 + e_0) \tag{9.10}$$

one obtains the "currents", which have the particularity to be isotropic,

$$j_L = \psi u e_0 \tilde{u} \tilde{\psi} = \psi u e_0 \tilde{\psi} = \frac{1}{2}\psi(e_0 - e_3)\tilde{\psi} = \frac{1}{2}(j - j') \tag{9.11}$$

$$j_R = \psi \tilde{u} e_0 u \tilde{\psi} = \psi \tilde{u} e_0 \tilde{\psi} = \frac{1}{2}\psi(e_0 + e_3)\tilde{\psi} = \frac{1}{2}(j + j') \tag{9.12}$$

9.2 Left and Right Doublets Associated with Two Wave Functions

We will only consider the case of the left doublets, the case of the right doublets giving similar results.

Let two ordinary Dirac spinors Ψ^1, Ψ^2 be, in their Hestenes form ψ^1, ψ^2, corresponding to two particles of spin 1/2.

What follows is applicable to all the invertible biquaternions. We define a left doublet ψ_L as

$$\psi_L = \psi_L^1 - \psi_L^2 e_1 \tag{9.13}$$

that we express in the form of a column $\psi_L \simeq \begin{pmatrix} \psi_L^1 \\ \psi_L^2 \end{pmatrix}$.

The choice of the vector $\mathbf{e}_1$ is arbitrary, all that we need is its orthogonality to $\mathbf{e}_3$ (see Nota below).

Each ψ_L^1, ψ_L^2 verifies the relation

$$\psi_L^\alpha \mathbf{e}_3 = -\psi_L^\alpha, \quad (\alpha = 1, 2) \tag{9.14}$$

One can write, because $\mathbf{e}_1^2 = 1$,

$$-\psi_L \mathbf{e}_1 = \psi^2 - \psi^1 \mathbf{e}_1 \quad \Leftrightarrow \quad \begin{pmatrix} \psi^2 \\ \psi^1 \end{pmatrix} = \begin{pmatrix} 0 & 1 \\ 1 & 0 \end{pmatrix} \begin{pmatrix} \psi^1 \\ \psi^2 \end{pmatrix} = \tau_1 \Psi_L \tag{9.15}$$

since, furthermore, $\mathbf{e}_2 = \mathbf{e}_1 \mathbf{e}_3 \underline{i} = -\mathbf{e}_3 \underline{i} \mathbf{e}_1$, applying (9.14),

$$-\psi \mathbf{e}_2 = -\psi^2 \underline{i} - \psi^1 \underline{i} \mathbf{e}_1 \quad \Leftrightarrow \quad \begin{pmatrix} -\psi^2 \underline{i} \\ \psi^1 \underline{i} \end{pmatrix} = \begin{pmatrix} 0 & -\underline{i} \\ \underline{i} & 0 \end{pmatrix} \begin{pmatrix} \psi^1 \\ \psi^2 \end{pmatrix} = \tau_2 \Psi_L \tag{9.16}$$

and since $\mathbf{e}_1 \mathbf{e}_3 = -\mathbf{e}_3 \mathbf{e}_1$, applying (9.14) again,

$$-\psi_L \mathbf{e}_3 = \psi^1 + \psi^2 \mathbf{e}_1 \quad \Leftrightarrow \quad \begin{pmatrix} \psi^1 \\ -\psi^2 \end{pmatrix} = \begin{pmatrix} 1 & 0 \\ 0 & -1 \end{pmatrix} \begin{pmatrix} \psi^1 \\ \psi^2 \end{pmatrix} = \tau_3 \Psi_L \tag{9.17}$$

So one can define three transformations, corresponding to the action of the τ_k matrices of the conventional presentation,

$$\psi_L \rightarrow -\psi_L \mathbf{e}_1, -\psi_L \mathbf{e}_2, -\psi_L \mathbf{e}_3 \quad \Leftrightarrow \quad \Psi_L \rightarrow \tau_1 \Psi_L, \tau_2 \Psi_L, \tau_3 \Psi_L \tag{9.18}$$

Assuming that the biquaternion ψ_L is invertible, that is $\psi_L \tilde{\psi}_L \neq 0$ (see a condition below), we deduce the correspondence similar to Eq. 8.7, except for a change of sign,

$$\bar{\Psi}_L \gamma^\mu \tau_k \Psi_L \Leftrightarrow -e^\mu \cdot (\psi_L \mathbf{e}_k \mathbf{e}_0 \tilde{\psi}_L) = -e^\mu \cdot (\psi_L \mathbf{e}_k \tilde{\psi}_L) = -j_k^\mu \tag{9.19}$$

Nota. The presence of $\mathbf{e}_1$ in the biquaternion ψ_L requires an explanation. The direction of $\mathbf{e}_3$ allows one to define at each point x of M, the directions of the "spin plane" of the electron and of the neutrino (by means of the orthogonality to the vector $R_e \mathbf{e}_3 R_e^{-1}$ and the vector $R_\nu \mathbf{e}_3 R_\nu^{-1}$). The choice of the direction of $\mathbf{e}_1$ is arbitrary for defining the direction of this plane for the two particles considered independently. But a commun choice is a necessity to obtain the correlation of the effect of a $U(1)$ change of gauge which is a rotation through a same angle χ of these planes upon themselves [1].

The choice of $\mathbf{e}_1$ allows also the translation to STA of the above isospin matrices. The choice of $\mathbf{e}_2$ would be in the same way acceptable but would correspond to a different but similar form of these matrices.

9.3 The Part $SU(2)$ of the $SU(2) \times U(1)$ Gauge

The biquaternion ψ_L is invertible if

$$\psi_L \tilde{\psi}_L = -(X + \tilde{X}) \neq 0, \quad X = \psi^2 u \mathbf{e}_1 \tilde{\psi}^1 \tag{9.20}$$

which is verified if X is not reduced to a bivector, that we will suppose.

Then we can consider the four currents which are not isotropic

$$j_L^0 = \psi_L e_0 \tilde{\psi}_L, \quad j_L^k = -\psi_L \mathbf{e}_k \tilde{\psi}_L \tag{9.21}$$

They give an orientation to the orthonormal frame $\{j_L^\mu\}$ negative with respect to the one of the galilean frame $\{e^\mu\}$, but do not change the properties of the $SU(2)$ gauge as they have been described in Sect. 8.

9.4 The Part $U(1)$ of the $SU(2) \times U(1)$ Gauge

A $U(1)$ change of gauge for the doublet is defined in STA by the transforms

$$\psi^1 \rightarrow \psi^1 e^{(\pm e_1 e_2 \varphi/2)}, \quad \psi^2 \rightarrow \psi^2 e^{(\pm e_1 e_2 \varphi/2)} \tag{9.22}$$

Note that the angle φ defining this change must be the same for ψ^1 and ψ^2.

It is possible to deduce from

$$e^{(\pm e_1 e_2 \varphi/2)} e_j e^{(\mp e_1 e_2 \varphi/2)} = e_j, \quad j = 0, 3 \tag{9.23}$$

that the product of $SU(2)$ by $U(1)$ is direct.

A precise verification of this property will be achieved in Sect. 10.3.

9.5 Geometrical Interpretation of the $SU(2) \times U(1)$ Gauge of a Left or Right Doublet

Properties similar to the ones of the left doublet may be established for a right doublet.

We deduce from all that precedes the following properties:

1. The $SU(2)XU(1)$ gauge may be applied *only* to a left or right doublet.
2. A change of such a product of gauges corresponds to a finite rotation in the three space orthogonal to the timelike current of the doublet and a finite rotation of a *same* angle φ in the "spin planes" of the particles whose wave functions are ψ^1 and ψ^2.

9.6 The Lagrangian in the $SU(2) \times U(1)$ Gauge

Since ψ_L is invertible the langrangian has the same form as in Eq. 8.21, except that, since the three-space $\{-j_1, -j_2, -j_3\}$ has an orientation inverse of $\{j_1, j_2, j_3\}$ and so $S(n)$ is to be replaced by $-S(n)$ [see Eq. 16.13] in Eq. 8.18 and L_I by $-L_I$ in Eq. 8.19 in such a way that L is changed into

$$L = L_I - gL_{II} \Leftrightarrow L = -e^\mu \cdot \left[\partial_\mu \psi_L \underline{i} e_0 \tilde{\psi}_L \right]_V - \frac{g}{2} W^k \cdot j_k \qquad (9.24)$$

Taking account this change, all that we have established in Sect. 8.2 is still applicable to the $SU(2) \times U(1)$ gauge of the weak theory.

References

1. R. Boudet, in *The Theory of the Electron*, ed. by J. Keller, Z. Oziewicz (Mexico, 1997), p. 321
2. F. Halzen, D. Martin, *Quarks and Leptons* (Wiley, USA, 1984)

Part VI
The Glashow–Salam–Weinberg Electroweak Theory

Chapter 10
The Electroweak Theory in STA: Global Presentation

Abstract Since all the necessary equivalences between complex and real language have been clarified, the electroweak theory may now be exposed only by means of the use of the STA. This Section relates to the enumeration of the entities implied in the theory.

Keywords Electron · Neutrino · Doublet · Singlets · Weinberg angle

10.1 General Approach

A presentation of the electroweak theory in STA has been achieved by Hestenes (see [1–3]). Our approach, achieved quite independently, is different and longer because we have used in fact a step by step translation of the standard presentation (see in particular [4] and [5]) into the STA.

The presentation will only concern the leptonic part of the theory in the first generation (electron and neutrino). The extension to the part of the theory which takes into account the hadronic currents, associated with the quarks "strange" and "down", in the "mixture of Cabibbo", and "up", does not present additional difficulties (see [4], p. 154–157).

As in [6], we have chosen to present the theory in the simplest form but with the conservation of its fundamental features.

In particular we will not mention the role of entities as hypercharges, certainly important, but they do not appear in the final results.

As explained in the Abstract we will simply present the theory with the use of the STA. The agreement with the standard presentation will be insured by the coincidence of the entities and equations independent of all galilean frame in the two mathematical, complex and real, approaches.

R. Boudet, *Quantum Mechanics in the Geometry of Space–Time*,
SpringerBriefs in Physics, DOI: 10.1007/978-3-642-19199-2_10,
© Roger Boudet 2011

10.2 The Particles and Their Wave Functions

Only two leptons are to be considered, the electron and the neutrino.

They will be considered in the "spin up" state.

Their wave functions will be expressed by the Hestenes spinors ψ^e and ψ^ν in a galilean frame $\{e_\mu\}$.

10.2.1 The Right and Left Parts of the Wave Functions of the Neutrino and the Electron

Four wave functions, which may be deduced from these two waves, are considered.

Applying the decomposition Eq. 9.6 to the neutrino and the electron (see [4], p. 147), we consider the wave functions

$$\psi^\alpha = \psi^\alpha u + \psi^\alpha \tilde{u} = \psi_L^\alpha + \psi_R^\alpha \quad (\alpha = \nu, e) \tag{10.1}$$

with, we recall, $u = (1 - e_3)/2$, $\tilde{u} = (1 + e_3)/2$, and we define the following wave functions

$$\psi_R^\nu = \psi^\nu \tilde{u}, \quad \psi_R^e = \psi^e \tilde{u} \tag{10.2}$$

$$\psi_L^\nu = \psi^\nu u, \quad \psi_L^e = \psi^e u \tag{10.3}$$

10.2.2 A Left Doublet and Two Singlets

Three wave functions are used.

The wave function of the left doublet of the theory is defined with the help of the left part of the electron and the left part of the neutrino (see [4], p. 147)

$$\psi_L = \psi_L^\nu - \psi_L^e e_1 = \psi^\nu u - \psi^e u e_1, \quad e_1 = e_1 e_0 \tag{10.4}$$

The two singlets are ψ_R^ν and ψ_R^e.

10.3 The Currents Associated with the Wave Functions

Some of the following currents appear directly in the lagrangian. Others may be considered as auxiliary.

10.3.1 The Current Associated with the Right and Left Parts of the Electron and Neutrino

Using Eqs. 9.12, 9.11 one has for the electron

$$j_R^e = \frac{1}{2}(j_e + j_e'), \quad j_L^e = \frac{1}{2}(j_e - j_e') \tag{10.5}$$

that is two isotropic currents, from which one deduces

$$j_R^e + j_L^e = j_e \tag{10.6}$$

which is a timelike vector as in the Dirac theory of the electron.

In the same way one has for the neutrino

$$j_R^\nu = \frac{1}{2}(j_\nu + j_\nu'), \quad j_L^\nu = \frac{1}{2}(j_\nu - j_\nu'), \quad j_R^\nu + j_L^\nu = j_\nu \tag{10.7}$$

10.3.2 The Currents Associated with the Left Doublet

These currents are those of Sect. 9 Eq. 9.21

$$j_L^0 = \psi_L e_0 \bar{\psi}_L, \quad j_L^k = -\psi_L e_k \bar{\psi}_L \tag{10.8}$$

which are not isotropic if Eq. 9.20 is satisfied and are defined by

$$j_L^0 = (\psi^\nu u - \psi^e u e_1) e_0 (\tilde{u} \tilde{\psi}^\nu + e_1 \tilde{u} \tilde{\psi}^e) = \psi^\nu u e_0 \tilde{\psi}^\nu + \psi^e u e_0 \tilde{\psi}^e$$

$$j_L^1 = -(\psi^\nu u - \psi^e u e_1) e_1 e_0 (\tilde{u} \tilde{\psi}^\nu + e_1 \tilde{u} \tilde{\psi}^e) = -(\psi^\nu u e_1 - \psi^e u) e_0 (\tilde{u} \tilde{\psi}^\nu + e_1 \tilde{u} \tilde{\psi}^e)$$

$$j_L^2 = -(\psi^\nu u - \psi^e u e_1) e_2 e_0 (\tilde{u} \tilde{\psi}^\nu + e_1 \tilde{u} \tilde{\psi}^e) = -\underline{i}[(\psi^\nu u e_1 - \psi^e u) e_3 (\tilde{u} \tilde{\psi}^\nu + e_1 \tilde{u} \tilde{\psi}^e)]$$

$$j_L^3 = -(\psi^\nu u - \psi^e u e_1) e_3 (\tilde{u} \tilde{\psi}^\nu + e_1 \tilde{u} \tilde{\psi}^e) = -\psi^\nu u e_3 \tilde{\psi}^\nu + \psi^e u e_3 \tilde{\psi}^e$$

Recalling that

$$e_k = e_k e_0, \quad k = 1, 2 \implies e_k u = u e_k, \quad e_k \tilde{u} = \tilde{u} e_k$$

and using Eqs. 9.8, 9.10, 9.11, we can write

$$j_L^0 = \frac{1}{2}(j_\nu - j_\nu' + j_e - j_e') \tag{10.9}$$

$$j_L^1 = \frac{1}{2}[\psi^e(e_0 - e_3)\tilde{\psi}^\nu + \psi^\nu(e_0 - e_3)\tilde{\psi}^e] \tag{10.10}$$

$$j_L^2 = \frac{1}{2}i[-\psi^e(e_0 - e_3)\tilde{\psi}^\nu + \psi^\nu(e_0 - e_3)\tilde{\psi}^e] \tag{10.11}$$

$$j_L^3 = \frac{1}{2}(j_\nu - j_\nu' - (j_e - j_e')) \tag{10.12}$$

For the proof concerning j_L^2 we have used

$$\mathbf{e}_2 = \underline{i}\mathbf{e}_1\mathbf{e}_3, \quad Q\underline{i} = \underline{i}Q, \quad \forall Q \in Cl^+(M)$$

Using Eq. 9.23 one can immediately deduce from these expressions of the currents that a change of gauge $U(1)$, that is a rotation of the same angle in the spin planes of the electron and the neutrino, corresponding to the transforms of ψ^α, $\alpha = e, \nu$, into $\psi^\alpha \exp(\pm e_1 e_2 \varphi/2)$, leaves the j_L^μ unchanged and so does not affect a $SU(2)$ change of gauge, that is a rotation in the three-space orthogonal to j_L^0, which concerns the vectors j_L^k. So one verifies that the product of $SU(2)$ and $U(1)$ in the $SU(2) \times U(1)$ gauge concerning the left doublet is direct.

10.3.3 The Charge Currents

To be in agreement with the conventional presentation of the GSW theory, we introduce the so-called charged current

$$j_C = \frac{1}{2}\psi^e(e_0 - e_3)\tilde{\psi}^\nu \tag{10.13}$$

and its "complex conjugate" [see Eq. 10.15]

$$\tilde{j}_C = \frac{1}{2}\psi^\nu(e_0 - e_3)\tilde{\psi}^e \tag{10.14}$$

which are of the so-called V-A type, that is "vector-(axial-vector)" (see [4], p. 146), as that appears from Eqs. 10.15 and 10.16, like the sum of a vector and a pseudo-vector of M. Indeed one can easily check that if $Q_1, Q_2 \in C^+(M)$,

$$a \in M; \Rightarrow Q_1 a \tilde{Q}_2 = b + \underline{i}c$$

$$(Q_1 a \tilde{Q}_2)\tilde{} = Q_2 a \tilde{Q}_1 = b - \underline{i}c, \quad b, c \in M \tag{10.15}$$

and by another way one can deduce from Eqs. 10.10 and 10.11

$$jc = \frac{1}{2}(j_L^1 + \underline{i}j_L^2), \quad \tilde{j}c = \frac{1}{2}(j_L^1 - \underline{i}j_L^2) \qquad (10.16)$$

which in agreement with the property (10.15).

Then one can write

$$j_L^1 = jc + \tilde{j}c, \quad j_L^2 = \underline{i}(-jc + \tilde{j}c) \qquad (10.17)$$

10.4 The Bosons and the Physical Constants

10.4.1 The Physical Constants

The fundamental physical constants used are c, $\hbar$, e, and the angle θ of Weinberg, or weak mixing angle, so that $\sin^2 \theta = 0.234$ in the standard model ([5], p. 296, 301).

Two other constants, called g, g' ([5], Eq. 13.23) and ([4], Eq. 7.50) are deduced from e and θ:

$$g \sin \theta = g' \cos \theta = e > 0 \qquad (10.18)$$

(for the choice $e > 0$ see Sect. 6.1 and [5], Eq. 5.17).

10.4.2 The Bosons

1. The electromagnetic potential $A \in M$.
2. Two massive charged bosons W^1, $W^2 \in M$.
3. Two neutral bosons W^3, $B \in M$ and $Z \in M$, which are massive, such that ([4], Eq. 7.41)

$$W^3 = \sin \theta\ A + \cos \theta\ Z, \quad B = \cos \theta\ A - \sin \theta\ Z, \qquad (10.19_1)$$

and in another but equivalent presentation ([5], Eqs. 13.19, 13.20)

$$A = \sin \theta\ W^3 + \cos \theta\ B, \quad Z = \cos \theta\ W^3 - \sin \theta\ B, \qquad (10.19_2)$$

such that W^3, B, θ make the combination $\sin \theta\ W^3 + \cos \theta\ B$ massless.

10.5 The Lagrangian

The Lagrangian is in the form

$$L = L_I - L_{II} \qquad (10.20)$$

It contains:

1. A part L_I independent of the bosons fields (see Sect. 11.1)

$$L_I = \hbar c(e^\mu.[(\partial_\mu \psi^\nu)\underline{i}e_3 \tilde{\psi}^\nu]_V + e^\mu.[(\partial_\mu \psi^e)\underline{i}e_3 \tilde{\psi}^e]_V) = L_\nu + L_e \quad (10.21)$$

in which L_ν and L_e are nothing else but the traces of the momentum energy tensors of the neutrino and the electron (minus the term containing the potential).In particular L_e lies in the lagrangian Eq. 6.16 of the Dirac equation of the electron.

2. A part L_{II} implying the bosons fields

$$L_{II} = \frac{g}{2}(W^1.j_L^1 + W^2.j_L^2 + W^3.j_L^3) - g'B.\left(\frac{1}{2}j_L^0 + j_R^e\right) \quad (10.22)$$

L_I and L_{II} are each cut into several parts with a correspondence between each part of L_I with a part of L_{II}.

References

1. D. Hestenes, Space–time structure of weak and electromagnetic interactions. Found. Phys. **12**, 153–168 (1982)
2. D. Hestenes, Ann. J. Math. Phys. **71**, 718 (2003)
3. D. Hestenes, Am. J. Phys. **71**, 631 (2003)
4. E. Elbaz, *De l'électromagnétisme à l'électrobaible* (Ed. Marketing, Paris, 1989)
5. F. Halzen, D. Martin, *Quarks and Leptons*. (Wiley, USA, 1984)
6. R. Boudet, in *The Theory of the Electron,* ed. by J. Keller, Z. Oziewicz, eds (Mexico, 1997), p. 321

Chapter 11
The Electroweak Theory in STA: Local Presentation

Abstract This section is relative to the standard decompositions of the theory into several parts and the physical meaning of each of its part.

Keywords Charged · Neutral contribution · U(1) · SU(2) × (U1)

11.1 The Two Equivalent Decompositions of the Part L_I of the Lagrangian

One considers:

1. The decomposition Eq. 10.21

$$L_I = L_v + L_e \tag{11.1}$$

2. Another one (see [1], Eq. 7.20)

$$L_I = L_{I,L} + L^v_{I,R} + L^e_{I,R} \tag{11.2}$$

whose terms correspond to the left doublet and the two right singlets.

The definition of $L_{I,L}$ is deduced from Eqs. 3.6, 3.5 and a correspondence analog to Eq. (B.15)

$$\overline{\Psi}_L \gamma^\mu \partial_\mu \Psi_L \Leftrightarrow [e^\mu ((\partial_\mu \psi_L) \underline{i} e_3) e_0 \tilde{\psi}_L]_S = -[e^\mu (\partial_\mu \psi_L) \underline{i} e_0 \tilde{\psi}_L]_S \tag{11.3}$$

where $u e_3 = -u$ has been used.

From $\tilde{e}_1 = -e_1$ and properties Eqs. 9.5 to 9.10 one deduces from Eqs. 10.3, 10.4

$$L_{I,L} = \hbar c e^\mu \cdot \left[\frac{1}{2}((\partial_\mu \psi^v) \underline{i}(e_3 - e_0) \tilde{\psi}^v) + \frac{1}{2}((\partial_\mu \psi^e) \underline{i}(e_3 - e_0) \tilde{\psi}^e) \right]_V \tag{11.4}$$

R. Boudet, *Quantum Mechanics in the Geometry of Space–Time*,
SpringerBriefs in Physics, DOI: 10.1007/978-3-642-19199-2_11,
© Roger Boudet 2011

Using again Eqs. 3.6, 3.5 one deduces from Eqs. 10.2, 10.4

$$L^v_{I,R} = \hbar c [e^\mu (\partial_\mu \psi^v) \underline{i} e_3 \tilde{u} e_0 u \psi^v]_S \tag{11.5a}$$

$$L^e_{I,R} = \hbar c [e^\mu (\partial_\mu \psi^e) \underline{i} e_3 \tilde{u} e_0 u \tilde{\psi}^e]_S \tag{11.5b}$$

and since $e_3 \tilde{u} = \tilde{u}$, $\tilde{u} e_0 u = \tilde{u} e_0$

$$L^v_{I,R} = \hbar c e^\mu \cdot \left[\frac{1}{2} ((\partial_\mu \psi^v) \underline{i} (e_3 + e_0) \tilde{\psi}^v) \right]_V \tag{11.6}$$

$$L^e_{I,R} = \hbar c e^\mu \cdot \left[\frac{1}{2} ((\partial_\mu \psi^e) \underline{i} (e_3 + e_0) \tilde{\psi}^e) \right]_V \tag{11.7}$$

and so, after eliminations

$$L_I = \hbar c e^\mu \cdot [(\partial_\mu \psi^v) \underline{i} e_3 \tilde{\psi}^v]_V + e^\mu \cdot [(\partial_\mu \psi^e) \underline{i} e_3 \tilde{\psi}^e]_V = L_v + L_e \tag{11.8}$$

in conformity with Eq. 10.21.

11.2 The Decomposition of the Part L_{II} of the Lagrangian into a Charged and a Neutral Contribution

The following decomposition in two parts of L_{II}

$$L_{II} = L_{II,C} + L_{II,N} \tag{11.9}$$

implies a surprising particularity, that is a double role to the boson W^3. On one side it is related with W^1 and W^2 in such a way that the use of the gauge $SU2 \times U(1)$ appears in the lagrangian, and on the other W^3, which appears in the first part, is cut from these two bosons for being implied in the second part.

The contribution L to the lagrangian of the boson field is expressed into the sum $L_{II} = L_{I,C} + L_{I,N}$ where

$$L_{II,C} = \frac{g}{2} \left(W^1 \cdot j^1_L + W^2 \cdot j^2_L \right) \tag{11.10}$$

$$L_{II,N} = \frac{g}{2} W^3 \cdot j^3_L - g' B \cdot \left(\frac{1}{2} j^0_L + j^e_R \right) \tag{11.11}$$

(for the standard presentation see [1], Eqs. 7.29 and 7.40)

$L_{II,C}$ and $L_{II,N}$ are respectively called the charged and neutral contributions, because the boson gauge fields imply W^1, W^2 for $L_{II,C}$ and W^3, B (or A and Z by means of the Weinberg relations Eq. 10.19) for $L_{II,N}$.

11.2.1 The Charged Contribution

One deduces from Eqs. 10.17, 11.10,

$$L_{II,C} = \left[(W^1(jc + \tilde{j}c) + \frac{g}{2} \left[W^2 \underline{i}(-jc + \tilde{j}c) \right] \right]_s \qquad (11.12)$$

where $[X]_{sc}$ means the scalar part of X, or, recalling that $\underline{i}a = -a\underline{i}$ if $a \in M$,

$$L_{II,C} = \frac{g}{2}[(W^1 + \underline{i}W^2)jc + (W^1 - \underline{i}W^2)\tilde{j}c)]_s \qquad (11.13)$$

Introducing the "complex" (in fact real since $\underline{i}$ is real) vectorial boson gauge fields

$$W^\pm = \frac{1}{\sqrt{2}}(W^1 \mp \underline{i}W^2) \qquad (11.14)$$

one has

$$L_{II,C} = \frac{g}{\sqrt{2}}[W^- jc + W^+ \tilde{j}c]_s \qquad (11.15)$$

see [1] Eq. 7.30, 7.31, 7.35).

11.2.2 The Neutral Contribution

One considers the following decomposition of $L_{II,N}$

$$L_{II,N} = L^A_{C,N} + L^Z_{C,N} \qquad (11.16)$$

where $L^A_{C,N}$ and $L^Z_{C,N}$ imply the electromagnetic potential A and the Z boson gauge field, respectively.

Using $g = e/\sin\theta$, $g' = e/\cos\theta$ and Eq. 10.18 one can write

$$L^A_{II,N} = eA \cdot \left[\frac{1}{2}(j^3_L - j^0_L) - j^e_R \right]$$

and using Eqs. 10.12, 10.9, 10.6

$$L^A_{II,N} = -eA \cdot j_e = qA.j_e, \quad q = -e < 0 \qquad (11.17)$$

(as [1] Eq. 7.72) where $-ej_e$ is to be interpreted now as the current density of charge of the electron (as in [2], Eq. 5.17 with $e > 0$).
    Also

$$L^Z_{II,N} = Z \cdot \left[\frac{1}{2}(g\cos\theta j^3_L + g'\sin\theta j^0_L) + g'\sin\theta j_R \right]$$

and in the same way

$$L_{II,N}^Z = Z \cdot \left[\frac{1}{2}(g \cos \theta + g' \sin \theta) j_L^3 + g' \sin \theta j_e \right] \qquad (11.18)$$

One can write

$$g \cos \theta = \frac{e \cos^2 \theta}{\sin \theta \cos \theta}, \quad g' \sin \theta = \frac{e \sin^2 \theta}{\sin \theta \cos \theta}$$

We introduce the following "current" with the change in this equation of $-j_{em}$ into j_e (see [2], Eq. 13.10),

$$j_N = j_L^3 + 2 \sin^2 \theta j_e \qquad (11.19)$$

or (see Eq. 10.12)

$$j_N = \frac{1}{2}[\psi^\nu(e_0 - e_3)\tilde{\psi}^\nu - \psi^e(e_0 - e_3 - 4 \sin^2 \theta e_0)\tilde{\psi}^e] \qquad (11.19)'$$

This current j_N, STA form of j_{NC}, Eq. 7.67 of [1], is equal to $2j^{NC}$ where J^{NC}, Eq. 13.25 of [2], is deduced from Eqs. 13.1, 13.6.

We obtain the contribution of the Z boson field, Eq. 7.74 of [1]

$$L_{II,N}^Z = \frac{e}{2 \sin \theta \cos \theta} Z \cdot j_N \qquad (11.20)$$

11.3 The Gauges

Two gauges are present in the theory:

11.3.1 The Part $U(1)$ of the $SU(2) \times (U1)$ Gauge

We extract $\hbar c e^\mu \cdot [(\partial_\mu \psi^e)\underline{i} e_3 \tilde{\psi}^e]_V$ from L_I and $-qA.j_e$ from $-L_{II}$ and we have

$$L_{U(1)} = \hbar c e^\mu \cdot [(\partial_\mu \psi^e)\underline{i} e_3 \tilde{\psi}^e]_V - qA.j_e, \quad q = -e < 0 \qquad (11.21)$$

which is nothing else but the part of the lagrangian of the Dirac electron Eq. 6.16, which, the term $-mc^2 \cos \beta$ being omitted, represents the presence of the $U(1)$ gauge.

11.3.2 The Part SU(2) of the SU(2) × (U1) Gauge

Tacking $L_{I,L}$ in L_I and $g(W^k \cdot j_L^k)/2$ in L_{II}, we have

$$L_{SU(2)} = -\hbar c e^{\mu} \cdot \left[(\partial_{\mu}\psi_L)\underline{i}e_0\tilde{\psi}_L\right]_V - \frac{g}{2}\left(W^1 \cdot j_L^1 + W^2 \cdot j_L^2 + W^3 \cdot j_L^3\right) \quad (11.22)$$

in conformity with Eq. 9.24.

The geometrical properties of the changes in the gauges have been treated, for $U(1)$, and for the part $U(2)$ of $SU(2) \times U(1)$ and the relations concerning the bivector fields (see [1], Eq. 7.26), in Chaps. 4 and 8, respectively.

11.3.3 Zitterbewegung and Electroweak Currents in Dirac Theory

Incorporation of electroweak currents in Dirac theory raises questions about how that relates to *zitterbewegung*, another prominent feature of standard Dirac theory. This issue has been studied by Hestenes [3] who concludes that it suggests further modification of Dirac theory. We quote from his paper (with adjustments in notation to conform to the present book):

"The usual Dirac current is given by

$$J = \psi^e e_0 \tilde{\psi}^e = \psi^e \frac{1}{2}(e_0 - e_3)\tilde{\psi}^e + \psi^e \frac{1}{2}(e_0 + e_3)\tilde{\psi}^e \quad (11.23)$$

where the right side separates the contribution of left- and -right handed components. The charged and neutral weak currents are

$$J_- = \psi^e (e_0 - e_3)\tilde{\psi}^v =)\tilde{J}_+ \quad (11.24)$$

$$J_Z = \psi^v \frac{1}{2}(e_0 - e_3)\tilde{\psi}^v - \psi^e \frac{1}{2}(e_0 - e_3 - 4\sin^2\theta e_0)\tilde{\psi}^e \quad (11.25)$$

....

Having aligned the standard model with geometry of the Dirac equation, we notice that one prominent feature of Dirac theory is missing, namely, the *zitterbewegung* (zbw) of the electron. The zbw was discovered and given its name by Schroedinger in an analysis of free particles solutions of the Dirac equation [4]. It has since been recognized as a general feature of electron phase fluctuations and proposed as fundamental principe of QM ([5, 6]).

One reason that the significance of zbw has been consistently overlooked, especially in electroweak theory, is that the relevant observables are not among the so-called "bilinear covariants," from which observable currents are constructed in the standard model. I refer to the vector fields $\psi e_1 \tilde{\psi} = \rho n_1$ and $\psi e_2 \tilde{\psi} = \rho n_2$, identified as observables in $\psi e_{\mu} \tilde{\psi} = \rho n_{\mu}$. The standard model deals only with

$\psi e_0 \tilde{\psi} = \rho n_0$ and $\psi e_3 \tilde{\psi} = \rho n_3$, especially in combination to form chiral currents in Eqs. 11.23, 11.24, and 11.25.

One way to recognize the significance of the observables n_2 and n_1 is to note they rotate with twice the electron phase along streamlines of the conserved Dirac current. The rotation rate for a free electron is the $2m_e/\hbar \Leftrightarrow 1\,\mathrm{ZHz}$ (zettaHertz), the *zwb frequency* found by Schroedinger. This rate varies in the presence of interactions but still remains outside the range of direct observation.

Other features of the zbw may be detectable, however. In particular, it has often been suggested that the electron's magnetic moment is generated by a circulating charged current. That suggestion is elevated to a principle by replacing the charged Dirac current $e\psi e_0 \tilde{\psi}$ by the *zbw current* $e\psi (e_0 - e_2)\tilde{\psi} = e\rho(n_0 - n_2)$. Obviously, the zbw current is analogous to the left-handed chiral current $\psi^e \frac{1}{2}(e_0 - e_3)\tilde{\psi}^e$ with e_3 replaced by e_2. Equation 11.23 expresses the Dirac current as a sum of left- and right-handed chiral currents. Therefore, we can incorporate zbw into electroweak theory by dropping the right-handed current and replacing the left-handed current by the zbw current.

There is no need for a zbw analog to the right-handed chiral current. Looking over the standard model in the preceding section, it is evident that the right-handed current plays only a minor role. Its main function is to balance the left-handed current to produce the Dirac current, as shown in Eq. 11.23. The theory may be simplified considerably once that function is seen to be unnecessary."

Note. Multiplied by $1/2$, J_- and J_+ are identical to J_C and $\tilde{J}_C$ of our Eqs. 10.13 and 10.14. J_Z is identical to our STA translation j_N, Eq. $11.19'$, of the standard expression of the current j_{NC} of [1] (achieved later but independently of the work of D. Hestenes) and, multiplied by 2, the current J^{NC} of [2].

References

1. E. Elbaz, De l'électromagnétisme à l'électrobaible (ed. by Marketing Paris, 1989)
2. F. Halzen, D. Martin, Quarks and leptons. J. Wiley and Sons, U.S.A. (1984)
3. D. Hestenes, in *Proceedings of the Eleventh Marcel Grossmann, Meeting on General Relativity.* ed. by H. Kleinert, R.T. Jantzen, R. Ruffini (World Scientific, Singapore 2008), pp. 629–647
4. A. Schrödinger Sitzungb. Preuss. Akad. Wiss. Phys-Math. Kl. **24**, 418 (1930)
5. D. Hestenes, in *Clifford Algebras and Their Applications in Mathematical Physics*, ed. by J. Chisholm, A. Common (Reidel. Pub. Comp., Dordrecht, 1986), pp. 321–346
6. D. Hestenes, Am. J. Phys. **71**, 104 (2003)

Part VII
On a Change of SU(3) into Three SU(2) × U(1)

Chapter 12
On a Change of SU(3) into Three SU(2)× U(1)

Abstract In the use of the group $SU(3)$, the simple replacement of the eighth vector G^8 by $G^8/\sqrt{3}$ allows to replace this group by the direct product of three $SU(2) \times U(1)$. Also it allows the geometrical interpretation of the theories in which $SU(3)$ is used. A question may be posed: is this change suitable for the Quantum Chromodynamics Theory?

Keywords Gell-Mann matrices · Gluons · Quarks

12.1 The Lie Group SU(3)

The group implies the use of:

12.1.1 The Gell–Mann Matrices λ_a

$$\lambda_1 = \begin{pmatrix} 0 & 1 & 0 \\ 1 & 0 & 0 \\ 0 & 0 & 0 \end{pmatrix}, \quad \lambda_2 = \begin{pmatrix} 0 & -i & 0 \\ i & 0 & 0 \\ 0 & 0 & 0 \end{pmatrix}, \quad \lambda_3 = \begin{pmatrix} 1 & 0 & 0 \\ 0 & -1 & 0 \\ 0 & 0 & 0 \end{pmatrix} \tag{12.1}$$

$$\lambda_4 = \begin{pmatrix} 0 & 0 & -1 \\ 0 & 0 & 0 \\ 1 & 0 & 0 \end{pmatrix}, \quad \lambda_5 = \begin{pmatrix} 0 & 0 & -i \\ 0 & 0 & 0 \\ i & 0 & 0 \end{pmatrix} \tag{12.2}$$

$$\lambda_6 = \begin{pmatrix} 0 & 0 & 0 \\ 0 & 0 & 1 \\ 0 & 1 & 0 \end{pmatrix}, \quad \lambda_7 = \begin{pmatrix} 0 & 0 & 0 \\ 0 & 0 & -i \\ 0 & i & 0 \end{pmatrix} \tag{12.3}$$

R. Boudet, *Quantum Mechanics in the Geometry of Space–Time*,
SpringerBriefs in Physics, DOI: 10.1007/978-3-642-19199-2_12,
© Roger Boudet 2011

$$\lambda_8 = \frac{1}{\sqrt{3}} \begin{pmatrix} 1 & 0 & 0 \\ 0 & 1 & 0 \\ 0 & 0 & -2 \end{pmatrix} \qquad (12.4)$$

12.1.2 The Column Ψ on which the Gell–Mann Matrices Act

Let us denote

$$\Psi = \begin{pmatrix} \Psi_1 \\ \Psi_2 \\ \Psi_3 \end{pmatrix} \qquad (12.5)$$

this column.

12.1.3 Eight Vectors G^a

Each vector $G^a \in M$ is associated with the matrix λ_a in the product $G^a \lambda_a$.

12.1.4 A Lagrangian

We mention only the part of the Lagrangian of a theory in which $SU(3)$ is used for the gauges implied by the theory. It is in the form

$$L = L_I - g L_{II}, \quad L_I = \bar{\Psi}\gamma^\mu i \partial_\mu \Psi, \quad L_{II} = g\bar{\Psi}\gamma^\mu G^a_\mu \lambda_a \Psi \qquad (12.6)$$

with summation on $a = 1, \ldots, 8$ (see for the Quantum Chromodynamics Theory, [1], Eq. 14.28, [2], p. 139, [3], p. 266).

12.1.5 On the Algebraic Nature of the Ψ_k

In the QCT the Ψ_k correspond to the wave functions of particles of spin 1/2. So they are to be considered in this theory like Dirac spinors.

12.1.6 Comments

The couple (Ψ_1, Ψ_2) is manipulated by the matrices $(\lambda_1, \lambda_2, \lambda_3)$ identical to the matrices τ_k. So is defined the only sub-group $SU(2)$ of $SU(3)$ which allows one to consider a Y. M. field and a geometrical interpretation of the gauge associated with this sub-group.

This triplet of matrices gives a privileged role to the couple (Ψ_1, Ψ_2).

But given the incompatibilities indicated in Chap. 17 (see also Sect. 8.3), it seems that, if the Ψ_k correspond to the waves functions of particles of spin 1/2, as is the case in chromodynamics, the spinors Ψ_1, Ψ_2, must be bound in a right or a left doublet.

In the particular case of the couple (Ψ_1, Ψ_2), a $SU(2) \times U(1)$ gauge appears in the use of $SU(3)$, with a $U(1)$ gauge and a $SU(2)$ gauge-Y. M.-field which may be interpreted in the geometry of space–time, as is established in Chap. 3 and 9.

But concerning the couples (Ψ_1, Ψ_3), (Ψ_2, Ψ_3) we have not found a possible interpretation in the real geometry of the Minkowski space–time, by the use of $SU(3)$, with Ψ_k particles of spin 1/2. It is the reason why we will propose the replacement of $SU(3)$ by the direct product of three $SU(2) \times U(1)$.

Note that $SU(3)$ has been associated via the Gell–Mann matrices by Hestenes in [4] to bivectors of M. Here we keep the way consistent with the presentation we have made of $SU(2) \times U(1)$ in the GSW theory as a step by step translation in STA of the standard complex theory.

12.2 A Passage From SU(3) to Three SU(2)× U(1)

In [5] we have presented an arrangement of the λ_a matrices, giving the part L_{II} of the lagrangian in a form *strictly equivalent* to the one of Eq. 12.6, but modifying completely its interpretation.

We introduce the vector

$$\hat{G}^8 = \frac{1}{\sqrt{3}}G^8 \tag{12.7}$$

and, with the aim that the product $G^8\lambda_8$ remains unchanged, we consider the matrix

$$\hat{\lambda}_8 = \sqrt{3}\lambda_8 \tag{12.8}$$

in such a way that

$$G^8\lambda_8 = \hat{G}^8\hat{\lambda}_8 \tag{12.9}$$

Then nothing is changed in the value of the lagrangian except, as we are going to see, the possibility of an other interpretation of the role of the gauges in the theory.

Indeed we can write

$$\hat{\lambda}_8 = \hat{\lambda}_8^1 + \hat{\lambda}_8^2 \tag{12.10}$$

where

$$\hat{\lambda}_8^1 = \begin{pmatrix} 1 & 0 & 0 \\ 0 & 0 & 0 \\ 0 & 0 & -1 \end{pmatrix}, \quad \hat{\lambda}_8^2 = \begin{pmatrix} 0 & 0 & 0 \\ 0 & 1 & 0 \\ 0 & 0 & -1 \end{pmatrix} \tag{12.11}$$

Introducing the column couples

$$\Psi_{12} = \begin{pmatrix} \Psi_1 \\ \Psi_2 \\ 0 \end{pmatrix}, \quad \Psi_{13} = \begin{pmatrix} \Psi_1 \\ 0 \\ \Psi_3 \end{pmatrix}, \quad \Psi_{23} = \begin{pmatrix} 0 \\ \Psi_2 \\ \Psi_3 \end{pmatrix} \tag{12.12}$$

one sees that the sets of the matrices $(\lambda_4, \lambda_5, \hat{\lambda}_8^1)$ and $(\lambda_6, \lambda_7, \hat{\lambda}_8^2)$ may be considered as isospin matrices τ_k acting on the couples Ψ_{13} and Ψ_{23}, respectively.

Then L_{II} may be written *in a stictly equivalent way*, except for the replacement of G^8 by $\hat{G}^8 = G^8/\sqrt{3}$,

$$L_{II} = g \left(\bar{\Psi}_{12} \gamma^\mu G_\mu^{3k} \tau_k \Psi_{12} + \bar{\Psi}_{13} \gamma^\mu G_\mu^{2k} \tau_k \Psi_{13} + \bar{\Psi}_{23} \gamma^\mu G_\mu^{1k} \tau_k \Psi_{23} \right) \tag{12.13}$$

with

$$G_\mu^{31} = G_\mu^1, \quad G_\mu^{32} = G_\mu^2, \quad G_\mu^{33} = G_\mu^3 \tag{12.14}$$

$$G_\mu^{21} = G_\mu^4, \quad G_\mu^{22} = G_\mu^5, \quad G_\mu^{23} = \hat{G}_\mu^8 \tag{12.15}$$

$$G_\mu^{11} = G_\mu^6, \quad G_\mu^{12} = G_\mu^7, \quad G_\mu^{13} = \hat{G}_\mu^8 \tag{12.16}$$

Now L_{II} corresponds to the direct product of three $SU(2)$, and if the Ψ_{ab} are considered as right or left doublets, the theory would imply now the direct product of three $SU(2) \times U(1)$ instead of $SU(3)$.

But, given the form Eq. 12.13 of L_{II}, the privilege given to the set $(\lambda_1, \lambda_2, \lambda_3)$ in the fact they are identical to the τ_k matrices is destroyed. What we have said about the couple (Ψ_1, Ψ_2) concerns now the couples (Ψ_1, Ψ_3) and (Ψ_2, Ψ_3) which appear as symmetrical. Furthermore one of the eight vectors, $\hat{G}^8$, is used twice, and that lead to the presence of nine vectors instead of eight. But two of them are identical, so the presence of eight distinct vectors still remains.

Note that the symmetry of the presence of identical eighth vectors in the couples (Ψ_1, Ψ_3) and (Ψ_2, Ψ_3) would seem particularly convenient in QCT, for the symmetry of the links $u_1 - d$ and $u_2 - d$ for the proton, $d_1 - u, d_2 - u$ for the neutron where u and d are the quarks "up" and "down".

However, this point of view requires a suitable definition of the coupling constant, because it is necessary to take into account in L_I the relation of Ψ_1 with Ψ_2 *and* Ψ_3 and the relation of Ψ_2 with Ψ_1 *and* Ψ_3. That may be made, without changing L, by considering that the coupling constant is $\hat{g} = g/2$ in such a way that the relative part of L_I with respect to L_{II} is multiplied by two. In a change of the gauge, the three Lorentz rotations U^c, like the one U considered in Sect. 8.2, may be chosen independently.

12.3 An Alternative to the Use of SU(3) in Quantum Chromodynamics Theory?

If the vectors G^a are called gluons and the Ψ_k are wave functions of quarks, all the definitions of Sect. 12.1 are inside the QCT, in its part concerning the gauge.

The alternative to the theory we have exposed in Sect. 12.2 is based by the replacement on G^8 by $G^8/\sqrt{3}$ and only experiments will be able to give an answer to the question of its suitability.

References

1. R. Boudet, in *Adv. appl. Clifford Alg* (Birkhaüser Verlag, Basel, Switzerland, 2008), p. 43
2. G. Chanfray, G. Smadja, Les particules et leurs symétries. (Masson, Paris, 1997)
3. E. Elbaz, De l'électromagnétisme à l'électrobaible. (Ed. Marketing, Paris, 1989)
4. F. Halzen, D. Martin, Quarks and Leptons. (Wiley, USA, 1984)
5. D. Hestenes, Space–time structure of weak and electromagnetic interactions. Found. Phys. **12**, 153–168 (1982)

Part VIII
Addendum

Chapter 13
A Real Quantum Electrodynamics

Abstract A construction of an electromagnetism which may be applied to charges as well classical as quantum, is proposed. Its applications to the radiations of light by hydrogenic atoms are recalled. Concerning the interaction between an electron with plane waves one shows that the complex Quantum Fields Theory may be replaced, in an equivalent way, by a pure real theory which avoids the recourse to unacceptable artifices.

Keywords Retarded potentials · Plane wave perturbation

13.1 General Approach

The Quantum Fields Theory (QFT), in which the electromagnetic potentials are "quantified", has been used in Quantum Mechanics from 1927 until now. Such a theory employs a complex language in apparent agreement with the language of complex matrices and spinors.

Its characteristic lies in the association "$i\hbar$" of the imaginary number $i = \sqrt{-1}$ with the reduced Planck constant $\hbar$, "*by exact analogy with the ordinary quantum theory*" ([1], Para. 7, p. 56, lines 10–11), present in particular in the Dirac theory of the electron.

It is indisputable that the use of this theory has leaded with success to the quasi-totality of the results in which the interaction between an electron and an electro-magnetic plane wave is to be considered.

If the QFT is not necessary for the calculation of the levels of energy of the electron in the Darwin solutions of the Dirac equation in hydrogenics atoms, this theory has allowed physicists to determine the slight shift (the Lamb shift) of each of these levels in the hydrogen atom. This result is considered as the best proof of the validity of the QFT.

R. Boudet, *Quantum Mechanics in the Geometry of Space–Time*,
SpringerBriefs in Physics, DOI: 10.1007/978-3-642-19199-2_13,
© Roger Boudet 2011

However some authors have used the so-called "semi-classical treatment of radiations" (see [2], Chaps. X, XIII) in which only the classical theory of the electromagnetism is used, in particular for the Lamb shift calculation [3].

In fact, one can find exactly the same results, at least concerning this calculation, with and without the use of QFT. (Compare [4], Eq. 37 and [5], Eq. 16).

A little group of physicists has considered that the QFT is not a good theory.

For our own part, we will say that the QFT is a theory mathematically irreproachable (Chap. 19), leading sometimes to simplifications by the employment of complex numbers, at other times to complications, but which may lead to the use of doubtful artifices as soon as a potential in the form q/r is to be used (see [6] and Sect. 19.3).

What follows is a construction of electromagnetism, using only the Grassmann algebra and the inner product of the Minkowski space–time M. The fundamental elements of this construction are as suitable to the quantum as the classical theory.

They are based in first place on the properties of the electromagnetic potential, because, in particular, it is the only electromagnetic entity to be considered in the Dirac theory in the determination of the levels of energy of the electron in the Darwin solution of the Dirac equation for the hydrogen-like atoms. It intervenes also in the Lamb shift calculation (see [1, 7] and Sect. 19.3).

13.2 Electromagnetism: The Electromagnetic Potential

13.2.1 Principles of the Potential

We propose four fundamental principles.

P1. All punctual charge q situated at a point P of the Minkowski space–time M is endowed with an unitary timelike vector v (such that $v^2 = 1$) called space–time velocity at P of the charge q.

P2. The electromagnetic action of this charge q on a charge q' situed at a point $X \in M$ is such that the vector $\overrightarrow{PX}$ is isotropic (so that $\overrightarrow{PX}^2 = 0$), P being situated in the past of X.

P3. This electromagnetic action is the vector $A(X)$, such that

$$A(X) = q\frac{v}{\overrightarrow{PX}.v} \qquad (13.1)$$

is called retarded potential created at X by the charge q situed at P.

Let us define

$$\overrightarrow{PX} = r(v+n), \quad \overrightarrow{PX} \cdot v = r > 0 \qquad (13.2)$$

$$n \cdot v = 0, \quad n^2 = -v^2 = -1, \quad (v+n) \cdot v = 1$$

we can write

$$A = q\frac{v}{r} \tag{13.3}$$

P4. The potential A created at a same point X by different charges q_i is the vector sum of their potentials A_i considered separately:

$$A(X) = \sum_i q_i \frac{v_i}{\overrightarrow{P_i X}.v_i} \tag{13.4}$$

Note that all the points P_i are all situated on the hyper-cone $H(X)$ whose top is X and that X cannot coincide with a point P_i.

Note. The expression (13.1) is analog to the Liénard and Wiechert potential which has been deduced in classical electromagnetism from the integral formula of the retarded potentials in the case where the charge is a small sphere whose center describes a straight line.

But here we only associate to the charge a scalar q, a point P of space–time, and an unit time-like vector v. That may be applied to the cases where:

1. In classical mechanics, the charge describes a trajectory whose tangent vector at P is v.
2. In quantum mechanics one considers that the presence of the charge q at the point P is only an eventuality, and that the charge does not have necessarily a trajectory.

13.2.2 The Potential Created by a Population of Charges

Let us consider the case of a numerous population of punctual charges q_i each one situated at a point P_i in a small neighbourhood ϖ of a point P in such a way that the vector v_i of each charge is about the same as an unic vector v.

We associate with P the total charge dq included in ϖ.

For applying the Principle P2, we have to take into account the inclusion of ϖ into the hypercone $H(X)$. That may be done by considering ϖ as generated by an isotropic vector

$$\xi = d\ell(v + n), \quad d\ell > 0$$

whose origin describes a portion of plane η orthogonal to v and n. If the size of η is small with respect to the length r, then the vector $\overrightarrow{P'X}$ may be considered as isotropic for any $P' \in \varpi$ and one can consider that $\varpi \subset H(X)$. The portion of plane η is situated inside in the three-space $E^3(v)$ orthogonal to v.

Note that the measure of the projection of ϖ upon $E^3(v)$ is $d\tau_0 = d\sigma d\ell$, where $d\sigma$ is the area of η, since $-\xi.n = d\ell$.

The potential $A(X)$ may be written in the form

$$A(X) = \sum_{P \in \Omega} dq\frac{v}{r} = \sum_{P \in \Omega} dq\frac{v}{\overrightarrow{PX}.v} \tag{13.5}$$

where Ω is the reunion of all the domains ϖ.

The point X cannot belong to the domain Ω. An exception is envisaged in the construction of the Maxwell laws, but is outside of the present study.

The above construction is applicable as well:

1. In classical electromagnetism, to a population of distinct charges.
2. In quantum electromagnetism, to the population of eventualities of presence of an unique charge. dq corresponds then to the product of a constant charge ($q = -e$ for the electron) by a local presence probability.

13.2.3 Notion of Charge Current

The charge density $j(P)$ associated with a domain of M centered in a point P is, at the limit where this domain is considered as infinitely small, the quotient of the total of the charges included in this domain, and the measure of the orthogonal projection of this domain to the three space $E^3(v)$ orthogonal to the common vector v of the charges. It is called a charges current.

Defining a charge density $\rho = dq/d\tau_0$, where dq is the charge contained in a small neighborhood ϖ of P, one can introduce the notion of charge current $j(P) = \rho v$ and use the formula (13.4) of the retarded potentials, but with the restrictive condition that, for a given measure $d\tau_0$ of the orthogonal projection of ϖ on $E^3(v)$, the shape of ϖ has no incidence (or a weak incidence) on the value of dq, and so that j is independent on the choice of the point X where the potential A is considered .

1. In classical electromagnetism, with a population of distinct charges, we will define

$$j = \rho v, \quad \rho = \frac{dq}{d\tau_0} \tag{13.6}$$

2. In quantum electromagnetism, concerning the population of eventualities of presence of an unique charge q ($q = -e < 0$ for the electron) we will define

$$j = q\rho_0 v, \quad dq = qdp, \quad \rho_0 = \frac{dp}{d\tau_0} > 0 \tag{13.7}$$

where dp corresponds to the local presence probability of the charge at the points $P_i \in \varpi$.

Note that, given its definition, the vector j is independent of all galilean frame.

In the use of the charge current in classical or quantum electromagnetism, two assumptions, confirmed by experimenst, are made, to give an answer to the two following questions.

1. The current which is defined in a point P is independent of all other point X, in a strict theory, following the above principles, this current could not be used for the determination of a potential $A(X)$.

2. It is not possible that there exists an accumulation of charges around a point P and that the current j obey the following property, called the conservation of the current:

$$\partial.j = 0, \quad \partial = e^\mu \partial_\mu \tag{13.8}$$

1. A way to elude the difficulty of the compatibility of the use of the current in the determination of the potential at a point X is to admit that the shape of ϖ has no incidence (or a weak incidence) on the value of dq, so that j is independent of the choice of the point X where the potential A is considered.
2. In other respects, one adopts the hypothesis that the shape of ϖ and the distribution of the charges around the point P are not incompatible with the property Eq. 13.8.

13.2.4 The Lorentz Formula of the Retarded Potentials

Taking into account the definition of the current, the fact, deduced from Eq. 13. 6 or 13. 7, that $vdq = j(P)d\tau_0$ and the two above hypothesis, we can write Eq. 13. 5 in the form

$$A(X) = \sum_{P \in \Omega} \frac{j(P)}{r} d\tau_0 = \sum_{P \in \Omega} \frac{j(P)}{\overrightarrow{PX}.v} d\tau_0 \tag{13.9}$$

the sign $\sum$ corresponding in fact to a summation on the volumes $\varpi, d\tau_0$ being defined as in Sect. 13.2.2 , multiplied by j/r.

One can express this formula, all of whose the terms of are invariant, in a galilean frame $\{e_\mu\}$.

We can write

$$v = \alpha e_0 + \beta N, \quad n = \beta e_0 + \alpha N, \quad N \cdot e_0 = 0, \quad N^2 = -1 \tag{13.10}$$

in such a way that $(v + n)^2 = 0$ is verified, then

$$\overrightarrow{PX} = r(v + n) = r(\alpha + \beta)(e_0 + N) = R(e_0 + N), \quad R = r(\alpha + \beta). \tag{13.11}$$

In the definition of the neighbourhood ϖ of P given in Sect. 13.2.2, the passage to the frame $\{e_\mu\}$ leaves the portion of plane η, which is inside the intersection of $E^3(v)$ and $E^3(e_0)$, unchanged, and in other respect we can write

$$d\ell(v + n) = d\ell(\alpha + \beta)(e_0 + N) = d\ell'(e_0 + N), \quad d\ell' = d\ell(\alpha + \beta)$$

and so we can define a portion of volume $d\tau'$ in the galilean frame $\{e_\mu\}$, corresponding to $d\tau_0$, such that

$$d\tau' = d\sigma d\ell' = d\sigma d\ell(\alpha + \beta) = d\tau_0(\alpha + \beta) \tag{13.12}$$

We deduce that

$$\frac{d\tau_0}{r} = \frac{d\tau'}{R} \tag{13.13}$$

and Eq. 13. 9 becomes

$$A(X) = \sum_{P \in \Omega} \frac{j(P)}{R} d\tau' \tag{13.14}$$

Let x^μ and a^μ of the frame $\{e_\mu\}$ be, so that

$$\overrightarrow{XP} = -\overrightarrow{PX} = (x^\mu - a^\mu)e_\mu, \quad x^0 - a^0 = -\overrightarrow{PX}.N > 0 \tag{13.15}$$

Eq. 13.14 may be written

$$A(x^0, x^k) = \int_V \frac{j(x^0 - R, a^k)}{R} d\tau' \tag{13.16}$$

$$R = x^0 - a^0 = \left[\sum_k (x^k - a^k)^2 \right]^{1/2} \tag{13.17}$$

where V is the volume in the space $E^3(e_0)$ which is the orthogonal projection upon $E^3(e_0)$ of the domain Ω of M containing all the charges which are the source of the potential $A(X)$ and $d\tau' = da^1 da^2 da^3$.

This equation is called the Lorentz formula of the retarded potentials.

Note that the equality $a^0 = x^0 - R$ contains an ambiguity which requires an explanation. With respect to the integral, a^0 is only the time coordinate of the point P and so is to be considered as independent of the x^μ and also the a^k though the a^k lie in the R of $x^0 - R$. But in the derivations with respect to the x^μ, and only in these derivations, a^0 is to be considered as a function of the x^μ.

13.2.5 On the Invariances in the Formula of the Retarded Potentials

In the integral (13.16), the vector $j(P)$ is independent of all galilean frame.

As an imperative necessity, the quotient $d\tau'/R$ must has this property of invariance.

In what precedes we have shown that this property is verified because $d\tau$ is deduced from the invariant $d\tau_0$ and also R is deduced from the invariant $\overrightarrow{PX}$, both in a projection on the three space $E^3(e_0)$ of the frame $\{e_\mu\}$.

However this $d\tau_0$ has been defined by means of a neighborhood of P included in the hyper-cone $H(X)$. Given the fact that R is related to the vector $\overrightarrow{PX}$, this definition of $d\tau_0$ appears as a necessity.

But the $d\tau_0$ which lies in the definition Eq. 13.6 of $j(P)$ is different. The way given in Sect. 13.2.3 to elude the difficulty presented by the divergence between the two definitions of $d\tau_0$ is a weak point of the formula of Lorentz.

Nevertheless this formula is in agreement with experiments and the approximations used are to be considered as satisfactory.

13.3 Electrodynamics: The Electromagnetic Field, the Lorentz Force

13.3.1 General Definition

1. The electromagnetic field is the bivector F

$$F = \partial \wedge A \in \wedge^2 M \tag{13.18}$$

2. The Lorentz force f is the action of an electromagnetic field F on a punctual charge q' situated at a point X, whose velocity at this point is the unit time-like vector v', such that

$$f = q' F \cdot v' = q'(\partial \wedge A) \cdot v' \in M \tag{13.19}$$

which has the dimension of a force.

13.3.2 Case of Two Punctual Charges: The Coulomb Law

13.3.2.1 Field of a Punctual Charge at Rest, or Coulomb Field

Suppose an unique charge q situated at a point P, at rest in a galilean frame whose spacetime vector is v (P describes then a straight line of M), and let $\{e_\mu\}$ be the orthonormal frame such that $e_0 = v$. Consider a point X such that

$$\overrightarrow{PX} = r(v + n), \quad \overrightarrow{PX} \cdot v = r > 0$$

If the point X is at rest in the frame, the time coordinate x^0 of X does not intervene and, since $e^k = -e_k$, ($k = 1, 2, 3$), the operator ∂ expressed in spherical coordinates is reduced to

$$\partial = -n\frac{d}{dr}, \quad \text{and } F = -qn \wedge v\frac{d}{dr}\left(\frac{1}{r}\right) = q\frac{n \wedge v}{r^2} \tag{13.20}$$

13.3.2.2 The Lorentz Force and the Coulomb Law

If two charges q, q', sitied at points P and X of the spacetime M, are at rest in a galilean frame whose vector time is $v \in M$, the force f acting on the charge q' is defined by the bivector $q'F$

$$q'F = q'q\frac{n \wedge v}{r^2}, \quad \overrightarrow{PX} = r(v + n), \quad v^2 = 1 = -n^2, \quad v \cdot n = 0 \quad (13.21)$$

where the vector $n \in M$ corresponds to the direction of the oriented straight line which joins the projections of P and X on the space $E(v)$.

Since the spacetime velocity of the charge q' is here v one has, using Eq. 2.1,

$$f = q'F \cdot v = q'q\frac{n}{r^2} \quad (13.22)$$

That is the Coulomb law which is the base of the electrostatic.

13.3.3 Electric and Magnetic Fields

The Electromagnetic bivector field $F = \partial \wedge A$ is independent of all galilean frame. But its effects are observed in experiments achieved in a particular galilean frame, "the laboratory frame" $\{e_\mu\}$.

In a galilean frame a bivector is split into two parts whose properties are geometrically different, one which implies e_0, the other which implies only the vectors e_k.

This geometrical difference is the reason for the separation of the field F into two parts, whose the physical properties, when they are observed in a laboratory galilean frame, are different.

Let us denote in conformity with the notations of Sect. 2.2

$$X = X^0 e_0 + \vec{X} \in M, \quad \vec{X} \wedge e_0 = \mathbf{X}, \quad \vec{X}_1 \wedge \vec{X}_2 = -\mathbf{X}_1 \hat{\wedge} \mathbf{X}_2 = -\underline{i}(\mathbf{X}_1 \times \mathbf{X}_2)$$

where $\mathbf{X}_2 \hat{\wedge} \mathbf{X}_1$ and $\mathbf{X}_2 \times \mathbf{X}_1$ mean here the Grassmann and cross products in $\mathcal{R}^{3,0}$, defining

$$a^k e_k = \vec{a}, \quad \vec{a} e_0 = \mathbf{a}, \quad \partial = e^\mu \partial_\mu = e^0 \partial_0 - \vec{\partial}, \quad \partial_\mu = \frac{\partial}{dx^\mu}$$

we can write

$$F = (e^0 \partial_0 - \vec{\partial}) \wedge (A^0 e_0 + \vec{A}) \Rightarrow F = \mathbf{E} + \underline{i}\mathbf{H} \quad (13.23)$$

with

$$\mathbf{E} = -\partial_0 \mathbf{A} - \nabla A^0, \quad \mathbf{H} = \nabla \times \mathbf{A}, \quad \partial_0 = \frac{\partial}{dx^0}, \quad \nabla = \mathbf{e}_k \frac{\partial}{dx^k} \quad (13.24)$$

13.3.4 Electric and Magnetic Fields Deduced from the Lorentz Potential

For the calculation of $\mathbf{E}$ and $\mathbf{H}$ it is convenient to express the Lorentz formula in $E^3(e_0)$. Denoting

$$\mathbf{r} = x^k \mathbf{e}_k, \quad \mathbf{r}' = a^k \mathbf{e}_k, \quad \mathbf{R} = \mathbf{r} - \mathbf{r}', \quad R = |\mathbf{R}|, \quad \mathbf{n} = \mathbf{R}/R, \quad \mathbf{n}^2 = 1$$

we deduce

$$A^0(x^0, \mathbf{r}) = \int_V \frac{j^0(x^0 - R, \mathbf{r}')}{R} \, d\tau', \tag{13.25}$$

$$\mathbf{A}(x^0, \mathbf{r}) = \int_V \frac{\mathbf{j}(x^0 - R, \mathbf{r}')}{R} \, d\tau', \tag{13.26}$$

Since $\vec{X}_1 \cdot \vec{X}_2 = -\mathbf{X}_1 \cdot \mathbf{X}_2$, and so $-\vec{\partial} \cdot \vec{X} = \nabla \cdot \mathbf{X}$, the equation of conservation takes the form

$$\partial_0 j^0 + \nabla \cdot \mathbf{j} = 0 \tag{13.27}$$

A complete construction of the two fields is pointed out by Krüger in [8] but we will consider here simply the case of the long range parts $\mathbf{E}^L(x^0, \mathbf{r})$ and $\mathbf{H}^L(x^0, \mathbf{r})$ of these fields, which correspond in general to the experimental observations. They are such that in the derivation of the integrals only the terms containing $1/R$ are taken into account, the terms containing $1/R^2$ being neglected (see [8]).

These two long range fields are in the form

$$\mathbf{E}^L(x^0, \mathbf{r}) = -\frac{\partial}{\partial x^0} \int_V \frac{\mathbf{j}^\perp(x^0 - R, \mathbf{r}')}{R} \, d\tau', \tag{13.28}$$

$$\mathbf{H}^L(x^0, \mathbf{r}) = -\frac{\partial}{\partial x^0} \int_V \frac{\mathbf{n} \times \mathbf{j}^\perp(x^0 - R, \mathbf{r}')}{R} \, d\tau' \tag{13.29}$$

The bivector $\mathbf{j}^\perp = \vec{j}^\perp \wedge e_0$ is so that $\vec{j}^\perp$ is the component of the spatial part $\vec{j}$ of the current vector j, orthogonal to the unit vector $\vec{n} = (x^k - a^k)e_k/R$ and is in the form $\mathbf{j}^\perp = \mathbf{j} - (\mathbf{j} \cdot \mathbf{n})\mathbf{n}$.

Note that the time component j^0 of the current does not intervene.

This forms Eqs 13.28 and 13.29 of $\mathbf{E}^L$ and $\mathbf{H}^L$ are particularly convenient for the calculation of the radiations in the hydrogenic atoms (see [9]).

We recall the proof of Eq. 28 (See [8]). One can write

$$-[\nabla A^0(x^0, \mathbf{r})]^L = -\int_V \frac{\nabla j^0(x^0 - R, \mathbf{r}')}{R} \, d\tau' \tag{13.30}$$

$$-\nabla j^0(x^0 - R, \mathbf{r}') = -\frac{\partial}{\partial a^0} j^0(a^0, \mathbf{r}') \frac{da^0}{dR}(\partial_k Re_k), \quad a^0 = x^0 - R$$

and since

$$\frac{da^0}{dR} = -1, \quad \partial_k Re_k = \frac{(x^k - a_k)e^k}{R} = \frac{\mathbf{R}}{R} = \mathbf{n} \tag{13.31}$$

$$-\nabla j^0(x^0 - R, \mathbf{r}') = \frac{\partial}{\partial a^0} j^0(a^0, \mathbf{r}')\mathbf{n}, \quad a^0 = x^0 - R \tag{13.32}$$

with

$$\frac{\partial}{\partial a^0} j^0(a^0, \mathbf{r}')\mathbf{n} = \frac{\partial}{\partial a^0} j^0(a^0, \mathbf{r}') \frac{da^0}{dx^0}\mathbf{n} = \frac{\partial}{\partial x^0} j^0(x^0 - R, \mathbf{r}')\mathbf{n}$$

We are going to use the following relation corresponding to the law of conservation of the current

$$\frac{\partial}{\partial x^0} j^0(x^0 - R, \mathbf{r}') = -\mathbf{e}_k \cdot \frac{\partial}{\partial x^k} \mathbf{j}(x^0 - R, \mathbf{r}') \tag{13.33}$$

Such a relation is ambiguous to the extend that the point where it is applied is the point $P \in V$ and the derivations are associated with the point X. But the fact that the time component a^0 of P is considered as a function $a^0 = (x^0 - R, \mathbf{r}')$ of the x^μ in $j^0(x^0 - R, \mathbf{r}')$ in the derivation Eq. 13.30 of the Lorentz formula Eq. 13.25 implies that a^0 is to be also considered as a function of the x^μ in the law of conservation and so in the vector $\mathbf{j}(x^0 - R, \mathbf{r}')$.

One can write

$$-\mathbf{e}_k \cdot \frac{\partial}{\partial x^k} \mathbf{j}(x^0 - R, \mathbf{r}') = -(\mathbf{e}_k \partial_k R) \cdot \frac{\partial}{\partial a^0} \mathbf{j}(a^0, \mathbf{r}') \frac{da^0}{dR} \tag{13.34}$$

and taking account Eq. 13. 31 and

$$\frac{\partial}{\partial a^0} \mathbf{j}(a^0, \mathbf{r}') = \frac{\partial}{\partial a^0} \mathbf{j}(a^0, \mathbf{r}') \frac{da^0}{dx^0} = \frac{\partial}{\partial x^0} \mathbf{j}(x^0 - R, \mathbf{r}') \tag{13.35}$$

one has

$$-\mathbf{e}_k \cdot \frac{\partial}{\partial x^k} \mathbf{j}(x^0 - R, \mathbf{r}') = \mathbf{n} \cdot \frac{\partial}{\partial x^0} \mathbf{j}(x^0 - R, \mathbf{r}') \tag{13.36}$$

and so

$$-\nabla j^0(x^0 - R, \mathbf{r}') = \frac{\partial}{\partial x^0}[(\mathbf{j}(x^0 - R, \mathbf{r}')) \cdot \mathbf{n})\mathbf{n}] \tag{13.37}$$

and we deduce Eq. 13. 28 from Eqs. 13. 24, 13. 36:

$$\mathbf{E}^L(x^0, \mathbf{r}) = -\frac{\partial}{\partial x^0} \int_V \frac{\mathbf{j}(x^0 - R, \mathbf{r}') - [\mathbf{j}(x^0 - R, \mathbf{r}') \cdot \mathbf{n})\mathbf{n}]}{R} \, d\tau' \tag{13.38}$$

For the calculation of $\mathbf{H}^L(x^0, x^k)$ it is sufficient to replace in Eq. (13.36) the scalar product by the vector product in such a way that

$$\mathbf{e}_k \times \frac{\partial}{\partial x^k}\mathbf{j}(x^0 - R, \mathbf{r}') = -\frac{\partial}{\partial x^0}(\mathbf{n} \times \mathbf{j}(x^0 - R, \mathbf{r}'))$$

One deduces Eq. 13. 29 immediately from this relation, after elimination of the component of $\mathbf{j}$ parallel to $\mathbf{n}$.

The agreement of Eqs. 13. 28, 13. 29 and experiments is confirmed with the observations of radiations of hydrogen-like atoms, emitted light, spontaneous emission and Zeemann effect (see [9], Chaps. 2, 6, 7, 14) .

13.3.5 The Poynting Vector

The Poynting vector is defined by the cross product

$$\mathbf{P} = \mathbf{E} \times \mathbf{H} \tag{13.39}$$

It is related to the electromagnetism energy in the following way (see [10], Chap. XXXII).

Let us consider an electromagnetism field stored up in a volume V limited by a surface S. The decreasing Φ during a time dt of the total energy localized in V during a time dt is equal to the flux of $\mathbf{R}$ through S, during the same time.

$$\Phi = \int_S (\mathbf{E} \times \mathbf{H}).\mathbf{n} \, d\sigma \tag{13.40}$$

This flux allows in particular the calculation of spontaneous emission (see [9], Sects 2.4 and 7.3).

13.4 Electrodynamics in the Dirac Theory of the Electron

We will consider here the case of an electron submitted to a central potential at rest, for which the Dirac probability current may be clearly defined. This case is relevant to the theory of the hydrogen-like atoms in which the kernel may be considered as punctual, that we have studied in [9] with the help of STA.

It is the reason why we have not evoked the Maxwell laws, which are not to be taken into consideration in this study.

13.4.1 The Dirac Probability Currents

We will express the $E^3(e_0)$ space where the kernel is at rest in spherical coordinates (r, θ, φ), with the use of $Cl(3, 0)$

$$\mathbf{u} = \cos \varphi \; \mathbf{e}_1 + \sin \varphi \; \mathbf{e}_2, \quad \mathbf{v} = -\sin \varphi \; \mathbf{e}_1 + \cos \varphi \; \mathbf{e}_2$$

13.4.2 Current Associated with a Level E of Energy

It is in the form (see [9], Sect. 4.2.5)

$$j^0 = a(r, \theta)e_0, \quad \mathbf{j} = b(r, \theta)\mathbf{v}$$

and so is independent of x^0 and verifies the law of conservation $\partial_\mu j^\mu = 0$.

13.4.2.1 Transition Current Between Two States

The transition current between two states, corresponding to the levels of energy E_1, E_2 is such that (see [9], App. 18)

$$\int j^0(\mathbf{r})d\tau = 0 \tag{13.41}$$

Then the current may be written in the form (see [9], Sect. 5.2)

$$\mathbf{j} = \cos \omega x^0 \; \mathbf{j}_1 + \sin \omega x^0 \; \mathbf{j}_2, \quad \omega = \frac{E_2 - E_1}{\hbar c}$$

$$\mathbf{j}_1 = \cos \epsilon \varphi \; \mathbf{j}_I + \sin \epsilon \varphi \; \mathbf{j}_{II}, \quad \mathbf{j}_2 = -\sin \epsilon \varphi \; \mathbf{j}_I + \cos \epsilon \varphi \; \mathbf{j}_{II} \tag{13.42}$$

where $\epsilon = -1, 0, 1$, and

$$\mathbf{j}_I = b(r, \theta) \; \mathbf{v}, \quad \mathbf{j}_{II} = a(r, \theta) \; \mathbf{u} + c(r, \theta) \; \mathbf{e}_3$$

This current verifies also the law of conservation $\partial_\mu j^\mu = 0$ (see [9], App. 19).

13.4.3 Emission of an Electromagnetic Field

Equations (13. 28) and (13. 29) show that if the current j is time-independent the long range part of the field is null.

As it is the case of the probability current of the state of a bound electron, that explains the reason why no electromagnetic field may be observed outside a passage from one state to another.

On the contrary, the transition current between two states depends on the time.

In the Darwin solution this current is time-periodic with a linear or a circular polarization which allows the observation of the transition (see [9], Chaps. 2 and 5).

13.4.4 Spontaneous Emission

We consider the flux Φ, per unit of time, through a sphere S of large radius, of the Poynting vector of the field, created by the transition current between two states, of an electron bound in an atom.

Let us consider Φ as averaged on a period $T = 2\pi\omega$ of the source current, supposed time-periodic and let us denote by

$$\langle X \rangle = \frac{1}{T} \int_0^T X \, dx^0$$

the average of X.

Because $\mathbf{E}$ and $\mathbf{H} = \mathbf{n} \times \mathbf{E}$ are orthogonal to $\mathbf{n}$ we can write for a sphere S of center 0 of radius R

$$\Phi = \frac{c}{4\pi} \int_{S_0} \langle (\mathbf{E} \times \mathbf{H}) \cdot \mathbf{n} \rangle \, R^2 d\sigma \tag{13.43}$$

then

$$\Phi = \frac{c}{4\pi} \int_{S_0} \langle \mathbf{E}^2 \rangle \, R^2 d\sigma \tag{13.44}$$

where S_0 is the sphere unity.

The flux Φ allows us to calculate the number of transitions per unit of time, in the phenomena called "spontaneous emission".

If we consider the energy E released at each transition, the ratio Φ/E gives the number of transitions per second.

The number of these transitions may be experimentally observed, and, for comparison, the theoretical calculation is interesting (see [9], Sects. 2.4 and 7.3, in particular for the transitions $2P1/2 - 1S1/2$ and $2P1/2 - 1S1/2$).

13.4.5 Interaction with a Plane Wave

The study of the passage from a level of energy to another one in a hydrogen-like atom due to the effect of an external monochromatic electromagnetic wave (a photon) is quite different from that of spontaneous emission which belongs to a simple statistical way. This effect is studied rather from a probabilistic point of view, in the frame of an approximation method.

We will recall the results we have given in Chap. 8, App. G and H of [9] with complement of proof for some of them.

13.4.5.1 An Approximation Method for Time-dependent Perturbation

We will follow the method of perturbation described in Sects. 29, 32 of the book of Schiff [2]. But here this method will be directly applied to the Dirac instead of the Schrödinger theory of the electron, and with the use of the real formalism. We recall the results established in App. G of [9].

Let us consider a wave function ψ in the form

$$\psi(x^0, \mathbf{r}) = \sum_n a_n(x^0)\psi_n(x^0, \mathbf{r}), \quad a_n(x^0) \in R \qquad (13.45)$$

where each ψ_n is the solution Eq. 4.5 of [9] for an electron in an hydrogenic atom in a state of energy E_n.

We suppose that, at a time $t = x_0/c$, a potential $A = A^k e_k$, written $\mathbf{A} = A^k \mathbf{e}_k$ in $Cl(E^3)$, is added to the central potential $A^0 e_0$ such that $eA^0 = V(r)$, ($e = -q > 0$).

Defining

$$\mathbf{g}_{mn} = \sin(\omega_{mn}x^0)\mathbf{j}_{1,mn} - \cos(\omega_{mn}x^0)\mathbf{j}_{2,mn}, \quad \omega_{mn} = \frac{E_n - E_m}{\hbar c} \qquad (13.46)$$

we have (see [9], Eq. G.7 to G.10)

$$\dot{a}_m(x^0) = \frac{e}{\hbar c} \int \mathbf{A} \cdot \sum_n a_n(x^0)\mathbf{g}_{mn}(x^0, \mathbf{r})d\tau \qquad (13.47)$$

The perturbation approximation method (see [2], Sect. 29) consists in replacing $\mathbf{A}$ by $\lambda\mathbf{A}$ and expressing each a_n as power series in λ :

$$a_n = a_n^{(0)} + \lambda a_n^{(1)} + \lambda^2 a_n^{(2)} + \cdots$$

Each term of the series corresponds to an order of approximation. We will consider only the first order, which consists in the calculation of $a_n^{(1)}$.

Equating the coefficients of equal power of λ we obtain

$$\dot{a}_m^{(1)}(x^0) = \frac{e}{\hbar c} \int \mathbf{A}(x^0, \mathbf{r}) . \sum_n c_n \mathbf{g}_{mn}(x^0, \mathbf{r}) d\tau$$

where c_n is a constant which may may be chosen as being δ_{mn}.

Now we consider two particular states j and k. The first one will be considered as the state of the electron before the beginning of the perturbation and the second one as the expected final state. So we have (see [9], Eq. G.15)

$$\dot{a}_j^{(1)}(x^0) = \frac{e}{\hbar c} \int [\mathbf{A}(x^0, \mathbf{r}) . (\sin \omega_{jk} x^0) \mathbf{j}_{1, jk}(\mathbf{r}) - \cos(\omega_{mn} x^0) \mathbf{j}_{2, jk}(\mathbf{r}))] d\tau \quad (13.48)$$

13.4.5.2 Perturbation by a Plane Wave

Part II of [9] has been devoted to the field created in a transition between two states corresponding to the levels of energies E_1, E_2, and the phenomena of spontaneous emission (in which the final level is lower), in the absence of all external action.

Now we are going to recall the principal results of Part III of [9] in which one takes into account the effect of a monochromatic electromagnetic wave with a propagation vector $\mathbf{k}$ of magnitude $2\pi \nu/c$ and a polarization whose direction, orthogonal to $\mathbf{k}$, will be represented by an unit vector $\mathbf{L}$.

When the light of quantum energy $h\nu$ falls on an electron, bound in an atom, whose energy is $E_1 > 0$, a quantum may be absorbed and the electron jumps into a state of energy $E_2 = E_1 + h\nu$. The energy E_1 belongs to the discrete spectrum and E_2 may belong to the discrete (bound–bound transition) or to the continuous spectrum (photoeffect).

In this case the potential $\mathbf{A}$ is such that

$$\mathbf{A} = eU \cos(\mathbf{k} \cdot \mathbf{r} - \varpi x^0 + \xi) \mathbf{L} \quad (13.49)$$

$$\varpi = \frac{2\pi \nu}{c}, \quad \mathbf{k} = \varpi \mathbf{K}, \quad \mathbf{K}^2 = \mathbf{L}^2 = 1, \quad \mathbf{K} \cdot \mathbf{L} = 0$$

where U is constant and ξ is a phase constant.

The way that we follow here differs partially from the one of [2] but leads to the same conclusion. It is applied here directly to the Dirac theory of the electron instead of the Schrödinger one.

We will denote now $j = 1, k = 2$ with $\omega_{21} = \omega = (E_2 - E_1)/\hbar c$.

A simple calculation shows that Eq. 13.48 becomes

$$\dot{a}_1^{(1)}(x^0) = \alpha U \mathbf{L} \cdot \left[\int \cos(\mathbf{k} \cdot \mathbf{r} - \varpi x^0 + \xi)(\sin(\omega x^0) \mathbf{j}_1^{\perp} - \cos(\omega x^0) \mathbf{j}_2^{\perp}) d\tau \right]$$

$$(13.50)$$

where $\alpha = e^2/\hbar c$ (e in e.s.u.) is the fine structure constant and $\mathbf{j}_k^{\perp}$ is the component of the vector $\mathbf{j}_k$ orthogonal to $\mathbf{k}$ because $\mathbf{L}$ being orthogonal to $\mathbf{k}$, the component of $\mathbf{j}_k(\mathbf{r})$ upon the direction of $\mathbf{k}$ does not intervene.

Now we define a vector which plays an important role in the theory of the perturbation by a plane wave (see Chaps. 9, 12, 13 of [9]), also in the Lamb shift calculation (Addendum of [9])

$$\mathbf{T}_j^{\perp}(\mathbf{k}) = \frac{1}{2} \int \cos(\mathbf{k.r}) \, \mathbf{j}_j^{\perp}(\mathbf{r}) \, d\tau \tag{13.51}$$

Note. The factor $1/2$ is not present in the usual presentation of these two vectors because it is present in the usual definition of the transition currents. We have introduced the factor $1/2$ in the above definition because it is absent in our definition of the current between two states (see [9], Chap. 5). The absence of the factor $1/2$ in the definition of the transition current appears as a necessity for the concordance of the theoritical calculation of spontaneous emission and the experimental results concerning this phenomena (see [9] Sect. 7.3).

Furthermore we will take into account the relation (see [9], Chap. 9, Eq. 9.12)

$$\int \sin(\mathbf{k} \cdot \mathbf{r}) d\tau = 0 \tag{13.52}$$

We are going to consider $\dot{a}_1^{(1)}(x^0)$ by developing Eq. 13.50 but with the omission of the terms implying $\sin(\mathbf{k} \cdot \mathbf{r})$, because of the relation Eq. 13.52, and also the terms implying $\omega + \varpi$ which are not be taken into account (see [2], Sect. 35) because the probability of finding the system in the state 2 after the perturbation requires that $\omega - \varpi$ is close to zero.

We will carry out the eliminations like for example

$$I = \cos(\mathbf{k} \cdot \mathbf{r} - (\varpi x^0 - \xi)) \sin \omega x^0$$

$$I = \cos \mathbf{k} \cdot \mathbf{r} \cos(\varpi x^0 - \xi) \sin \omega x^0 + \sin \mathbf{k} \cdot \mathbf{r}(\ldots)$$

$$I = \cos \mathbf{k} \cdot \mathbf{r}[(\cos \varpi x^0 \cos \xi + \sin \varpi x^0 \sin \xi) \sin \omega x^0] + \sin \mathbf{k} \cdot \mathbf{r}(\ldots)$$

$$\cos \varpi x^0 \sin \omega x^0 = \frac{1}{2}(\sin(\omega x^0 - \varpi x^0) + \sin(\omega x^0 + \varpi x^0)) = \frac{1}{2} \sin(\omega x^0 - \varpi) + \cdots$$

Taking into account these two omissions and Eq. 13. 51 we obtain

$$\dot{a}_1^{(1)}(x^0) = \alpha U \mathbf{L} \cdot [\sin(\Omega x^0 + \xi) \mathbf{T}_1^{\perp}(\mathbf{k}) - \cos(\Omega x^0 + \xi) \mathbf{T}_2^{\perp}(\mathbf{k})] \tag{13.53}$$

where $\Omega = \omega - \varpi$.

(*Note.* This equation is the same as Eq. H.3, App. H of [9] without the factor 2 which lies in (H.3) because, in this appendix, the factor $1/2$ has been omitted in the definition of $\mathbf{T}_j^{\perp}(\mathbf{k})$).

We consider the integration with respect to x^0:

$$a_1^{(1)}(x^0) = \int_0^{x^0} \dot{a}_1^{(1)}(x)dx \qquad (13.54)$$

which gives

$$a_1^{(1)}(x^0) = \frac{\alpha U}{\Omega} \mathbf{L} \cdot [(\cos \xi - \cos(\xi + \Omega x^0))\mathbf{T}_1^{\perp}(\mathbf{k}) + (\sin \xi - \sin(\xi + \Omega x^0))\mathbf{T}_2^{\perp}(\mathbf{k})] \qquad (13.55)$$

The average of $[a_1^{(1)}(x^0)]^2$ upon the phase factor ξ

$$\left\langle [a_1^{(1)}(x^0)]^2 \right\rangle = \frac{1}{2\pi} \int_0^{2\pi} [a_1^{(1)}(x^0)]^2 d\xi$$

leads to the formula

$$\left\langle [a_1^{(1)}(x^0)]^2 \right\rangle = \frac{\alpha^2 U^2}{2} ([\mathbf{L} \cdot \mathbf{T}_1^{\perp}(\mathbf{k})]^2 + [\mathbf{L} \cdot \mathbf{T}_2^{\perp}(\mathbf{k})]^2, \frac{\sin^2((\omega - \varpi)x^0)/2)}{((\omega - \varpi)/2)^2} \qquad (13.56)$$

which give the probability that a transition in which the system is left in a higher state ($E_2 \simeq E_1 + \varpi$) has taken place at the time x^0.

Otherwise the average $\langle\langle [a_1^{(1)}(x^0)]^2 \rangle\rangle$ of $[\mathbf{L} \cdot \mathbf{T}_j^{\perp}(\mathbf{k})]^2$ on all the directions of $\mathbf{L}$ may be calculated by denoting

$$[\mathbf{L} \cdot \mathbf{T}_j^{\perp}(\mathbf{k})]^2 = [\mathbf{T}_j^{\perp}(\mathbf{k})]^2 \cos^2 \eta_j$$

where η_j is the angle of $\mathbf{L}$ and the direction of $\mathbf{T}_j^{\perp}(\mathbf{k}$, and writing

$$\frac{1}{2\pi} \int_0^{2\pi} [\mathbf{L} \cdot \mathbf{T}_j^{\perp}(\mathbf{k})]^2 d\eta_j = [\mathbf{T}_j^{\perp}(\mathbf{k})]^2 \frac{1}{2\pi} \int_0^{2\pi} \cos^2 \eta_j d\eta_j = \frac{1}{2}[\mathbf{T}_j^{\perp}(\mathbf{k})]^2$$

we obtain

$$\langle\langle [a_1^{(1)}(x^0)]^2 \rangle\rangle = \frac{\alpha^2 U^2}{4} ([\mathbf{T}_1^{\perp}(\mathbf{k})]^2 + [\mathbf{T}_2^{\perp}(\mathbf{k})]^2) \frac{\sin^2((\omega - \varpi)x^0)/2)}{[(\omega - \varpi)/2]^2} \qquad (13.57)$$

These formulas are similar to the one of [2], Eq. 35.16. They show that the probability for the transition from the state of energy E_1 to the state of energy E_2 is maximum (see [2], Fig. 27, p. 198) when $\varpi = \omega$.

For the study of the transition probability between the two states one considers what is called the matrix element of the transition (see [11], Eq. 59.3) introduced by H. Hall in 1936

$$D_j^{\mathbf{k},\mathbf{L}} = \mathbf{L} \cdot \mathbf{T}_j^{\perp}(\mathbf{k}) \tag{13.58}$$

which is in a complex form in the standard presentation but, because of the relation Eq. 13.52, is real a number.

The matrix elements, in their above real form, are used for the calculation of the photoeffect ([9], Part IV), its inverse phenomena, the radiative recombination (the emission of a photon after the capture of an electron by a hydrogen-like atom ([9], Chap. 13) and the Lamb shift.

13.4.6 The Lamb Shift

We have said that the use of the QFT is not necessary for the calculation of the Lamb shift and that the real language is sufficient.

The Lamb shift calculation implies the sum of three terms, *Electrodynamics energy term* W_D, the *Electrostatic energy term* W_S, and a term *the mass renormalization term* W_M which does not imply the electromagnetism.

The real vectors $\mathbf{T}_j^{\perp}(\mathbf{k})$ intervene in W_D in the form

$$[\mathbf{T}_1^{\perp}(\mathbf{k})]^2 + [\mathbf{T}_2^{\perp}(\mathbf{k})]^2 \tag{13.59}$$

the other scalars of the term being real.

So the problem of the determination of the matrix elements, is exactly the same for a transition in general and for the term W_D of the Lamb shift.

Note that W_S, which contains a potential q/r, is also real. It is presented in QFT in a complex form, which implies for the writing of this potential the use of an artifice (see Chap. 19).

To sum up, the calculation of the Lamb shift may be expressed, in agreement with the QFT, entirely in the Real Quantum Electrodynamics (see [9], Addendum).

As an example one can verify (see [9], p. 105) the concordance of the values calculated in this way of the contribution to the shift of the state $1s$ of the hydrogen atom due to all the states $2p$ of the discrete spectrum (see [9], p. 105) with the QFT ones of Seke [12].

References

1. W. Heitler, *The Quantum Theory of Radiation* (Clarendon Press, Oxford, 1964)
2. L. Schiff, Quantum Mechanichs. (Mc Graw-Hill, New York, 1955)
3. A.O. Barut, J.F. Van Huele, Phys. Rev. A **32**, 3187 (1985)
4. B. Blaive, R. Boudet, Ann. Fond. L. de Broglie **14**, 147 (1989)
5. J. Seke Z. Phys. D29:1(1994)
6. R. Boudet *New Frontiers in Quantum Electrodynamics and Quantum.* A. O. Barut ed. (Plenum, New York, 1990), p. 443

7. M. Kroll, W. Lamb, Phys. Rev. **75**, 388 (1949)
8. H. Krüger Elektrodynalik, 1 IBSN, Universitat Kaiserslauter (1985)
9. R. Boudet, Relativistic Transitions in the Hydrogenic Atoms. (Springer, Berlin, 2009)
10. G. Bruhat, Electricité. (Masson, Paris, 1956)
11. H. Bethe, E. Salpeter, Quantum Mechanics for One or Two-Electrons Atoms. (Springer, Berlin, 1957)
12. J. Seke, Mod. Phys. Lett. B **7**, 1287 (1993)

Part IX
Appendices

Chapter 14
Real Algebras Associated with an Euclidean Space

Abstract A construction of the Clifford algebra CI(E) associated with an euclidean space E is proposed. It is based on the Clifford products aA and Aa of a vector a of E and all element A of the Grassmann algebra of E, which include the inner product of E, these products being taken as definitions. One deduces then that a(Ab) = (aA)b, and so that CI(E) is an associative algebra. Its remarkable relation with the group O(E) is emphasized.

Keywords Grassmann · Inner · Clifford products

14.1 The Grassmann (or Exterior) Algebra of $\mathcal{R}^n$

We recall that $\wedge \mathcal{R}^n$ is an associative algebra generated by $\mathcal{R}$ and the vectors of $\mathcal{R}^n$ such that the Grassmann product $a_1 \wedge a_2 \wedge \cdots \wedge a_p$ of vectors $a_k \in \mathcal{R}^n$ is null if and only if the a_k are linearily dependent. If this product is non null, it is called a simple (or decomposable) p-vector and has the geometrical meaning of a p-paralleloid (a parallelogram if $p = 2$). The linear combination of simple p-vectors is called a p-vector, and the set of the p-vectors is a sub-space of $\wedge \mathcal{R}^n$ denoted $\wedge^p \mathcal{R}^n$, with $\wedge^0 \mathcal{R}^n = \mathcal{R}, \wedge^1 \mathcal{R}^n = \mathcal{R}^n, \lambda \wedge A = A \wedge \lambda = \lambda A, \lambda \in \mathcal{R}$.

We recall that $\wedge \mathcal{R}^n$ is the direct product of the sub-spaces $\wedge^p \mathcal{R}^n (p = 0, 1, \ldots n)$, each one of dimension C_n^p, and so $\dim(\wedge \mathcal{R}^n) = 2^n$, and the relation

$$a \wedge A_p = (-1)^p A_p \wedge a, \quad a \in \mathcal{R}^n, \quad A_p \in \wedge^p \mathcal{R}^n \qquad (14.1)$$

14.2 The Inner Products of an Euclidean Space $E = \mathcal{R}^{q,n-q}$

We denote $a.b \in \mathcal{R}$ the scalar (or inner) product of $a, b \in E$ defined by the signature $(q, n - q)$ of E.

R. Boudet, *Quantum Mechanics in the Geometry of Space–Time*,
SpringerBriefs in Physics, DOI: 10.1007/978-3-642-19199-2_14,
© Roger Boudet 2011

The inner products $A_p \cdot a, a \cdot A_p$ of a p-vector A_p by a vector a of E correspond to the operation so-called (by the physicists) "contraction on the indices" and verify

$$a \cdot A_p = (-1)^{p+1} A_p \cdot a, \quad a \in E, \quad A_p \in \wedge^p E$$

$$a \cdot A_0 = -A_0 \cdot a = 0, \quad A_0 \in \mathcal{R}$$

The product $a \cdot A_p$ is defined by

$$a \cdot A_p = \sum_{k=1}^{p} = (-1)^{k+1} (a \cdot a_k)(a_1 \wedge \cdots a_{k-1} \wedge a_{k+1} \wedge \cdots \wedge a_p) \qquad (14.3)$$

This definition may be written in another presentation.

$$a \cdot (a_1 \wedge A_{p-1}) = (a \cdot a_1) A_{p-1} - a_1 \wedge (a \cdot A_{p-1}), \quad A_{p-1} = a_2 \wedge \cdots \wedge a_p$$

The sum A of diverse p-vectors allows to write the general formula (whose the equivalent plays an important role in the theory of the p-forms of a vector space)

$$a \cdot (b \wedge A) = (a \cdot b)A - b \wedge (a \cdot A), \quad a, b \in E, \quad A \in \wedge E \qquad (14.4)$$

One deduce easily from Eq. 14.3

$$b \cdot (a \cdot A_p) = -a \cdot (b \cdot A_p) \qquad (14.5)$$

Indeed we have

$$b \cdot (a \cdot A_p) = (a \cdot a_1)(b \cdot a_2)(a_3 \wedge \cdots) - (a \cdot a_2)(b \cdot a_1)(a_3 \wedge \cdots) + \cdots$$

and permuting a and b, we see that Eq. 14.4 is satisfied.

Other properties of the inner product may be found in [1].

Note. A p-vector is nothing else but what the physicists call "an antisymmetric tensor of rank p" which is expressed by means of the components on a frame of E of the vectors of E which define this p-vector.

For example a simple bivector $a \wedge b$ of E will be written $a^i b^j - b^i a^j$ in a orthonormal frame $\{e_k\}$ of E.

In what follows we shall express an invariant p-vector by the Grassmann product of the vectors which define it, independently of all frame.

14.3 The Clifford Algebra $Cl(E)$ Associated with an Euclidean Space $E = \mathcal{R}^{p,n-p}$

The algebra $Cl(E)$ is a real associative algebra generated by $\mathcal{R}$ and the vectors of E whose elements may be identified to the ones of $\wedge E$.

The Clifford product of two elements A, B of $Cl(E)$ is denoted AB and verifies the fundamental relation

$$a^2 = a \cdot a, \quad \forall a \in E \tag{14.6}$$

from which we deduce

$$(a+b)^2 = a^2 + ab + ba + b^2 = (a+b) \cdot (a+b) = a \cdot a + 2a \cdot b + b \cdot b$$

and so

$$a \cdot b = \frac{1}{2}(ab + ba)$$

Now

$$ab = \frac{1}{2}(ab + ba) + \frac{1}{2}(ab - ba)$$

and identifying $(ab - ba)/2$ to $a \wedge b$, a convention that nothing forbids, one can write

$$ab = a \cdot b + a \wedge b, \quad (a, b \in E)$$

in such a way that

$$a \cdot b = 0 \Rightarrow ab = a \wedge b = -b \wedge a = -ba$$

We only mention the properties we need

$$aA = a \cdot A + a \wedge A, \quad Aa = A \cdot a + A \wedge a, \quad a \in E, \quad A \in \wedge E \tag{14.7}$$

which generalizes the relation about ab, from which one deduces, taking into account Eqs. 14.2, 14.1

$$a \cdot A_p = \frac{1}{2}(aA_p + (-1)^{p+1}A_p a), \quad a \wedge A_p = \frac{1}{2}(aA_p + (-1)^p A_p a) \tag{14.8}$$

If p vectors $a_i \in E$ are orthogonal their Clifford product verifies

$$a_1 \ldots a_p = a_1 \wedge \cdots \wedge a_p, \quad (a_k \in E, \; a_i \cdot a_j = 0 \text{ if } i \neq j) \tag{14.9}$$

The even sub-algebra $Cl^+(E)$ of $Cl(E)$ is composed by the sums of scalars and elements $a_1 \ldots a_p$ such that p is even.

One can immediately deduce from Eq. 14.9 that, using an orthonormal frame of E, the corresponding frame of $Cl(E)$ may be identified to the frame of $\wedge E$ and that $\dim(Cl(E)) = \dim(\wedge E) = 2^n$, and $\dim(Cl^+(E)) = 2^{n-1}$.

One uses the following operation called "principal antiautomorphism", or also "reversion",

$$A \in Cl(E) \rightarrow \tilde{A} \in Cl(E) \quad \text{so that } (AB)^\sim = \tilde{B}\tilde{A} \tag{14.10}$$

$$\tilde{\lambda} = \lambda, \quad \tilde{a} = a, \quad \lambda \in \mathcal{R}, \quad a \in E$$

The elements of $Cl(E)$ which are the sum of scalars and elements in the form $a_1 \ldots a_p$ such that p is even, define an algebra called even sub-algebra $Cl^+(E)$ of $Cl(E)$ whose dimension is 2^{n-1}.

14.4 A Construction of the Clifford Algebra

As a complement of proof of the formulas used in [1] we have acheived in [2] a construction of $Cl(E)$ in which $Cl(E)$ is built directly on the elements of $\wedge E$ in association with the euclidean structure of E.

Considering Eq. 14.7 as a *definition* (not as a property) of the Clifford products aA, Aa, we obtain as a theorem

$$a(Ab) = (aA)b, \quad a, b \in E, \quad A \in \wedge E \tag{14.11}$$

which endows $\wedge E$, taking into account the signature of E and the subsequent inner product, with the structure of an associative algebra, that is the Clifford algebra $Cl(E)$.

This construction requires a calculation. The definition of $Cl(E)$ given by the Bourbaki school, based only on the signature of E, is simple. But the proof of a vector space isomorphism bewteen $Cl(E)$ and $\wedge E$ requires a longer calculation.

One of the advantages of our construction is the fact that it makes directly apparent the formulas Eqs. 14.7 whose use is important in the application of a $Cl(E)$ to physics.

Using Eq. 14.7 (we repeat considered like an axiom) we can write for $a, b \in E, A \in \wedge E$

$$a(Ab) = a \cdot (A \cdot b) + a \wedge (A \cdot b) + a \cdot (A \wedge b) + a \wedge A \wedge b \tag{14.12}$$

Now in what follows we suppose that $A \in \wedge^p E$. Then A will be sum of diverse p-vectors and so any element of $\wedge E$.

Indeed, using Eqs. 14.2, 14.4 and 14.5

$$a \cdot (A \cdot b) = (-1)^{p+1} a \cdot (b \cdot A) = (-1)^p b \cdot (a \cdot A) = (-1)^{2p}(a \cdot A) \cdot b$$

$$a \wedge (A \cdot b) + a \cdot (A \wedge b) = (-1)^{p+1} a \wedge (b \cdot A) + (-1)^p a \cdot (b \wedge A)$$

$$= (-1)^{p+1} a \wedge (b \cdot A) + (-1)^p (a \cdot b)A - (-1)^p b \wedge (a \cdot A)$$

$$= (-1)^p b \cdot (a \wedge A) - (-1)^p(-1)^{p-1}(a \cdot A) \wedge b$$

$$= (-1)^{2p}(a \wedge A) \cdot b + (a \cdot A) \wedge b$$

and we obtain

$$(a \cdot A) \cdot b + (a \cdot A) \wedge b + (a \wedge A) \cdot b + a \wedge A \wedge b = (aA)b \tag{14.13}$$

and so Eq. (14.11) is verified.

14.5 The Group $O(E)$ in $Cl(E)$

In a general way, in all $Cl(E)$, the Clifford product $a_1a_2 \ldots a_p$, where ($a_k \in E, a_k^2 \neq 0$), may be associated with an isometry in the space E and so $Cl(E)$ may replace, in a very simple way, the general theory of the representations of the orthogonal group $O(E)$ in complex spaces.

Consider the relation

$$y = -bxb, \quad b^2 = b \cdot b \neq 0, \quad b, x \in E \tag{14.14}$$

Let the decomposition $x = x^{\perp} + x^{\|}$ be, where $x^{\|}$ and $x^{\perp}$ are parallel and orthogonal to b, respectively. Equation. 2.5 allows one to write

$$y = b^2(x^{\perp} - x^{\|}) \in E$$

and we see that the transformation $x \in E \rightarrow y \in E$ is a symmetry with respect to the hyperplan orthogonal to b, followed by the multiplication by the scalar b^2. The relation $z = -aya = abxba$ is a rotation followed by the multiplication by the scalar a^2b^2.

So the relation

$$y = (-1)^p I_p x \tilde{I}_p, \quad I_p = a_1a_2 \ldots a_p, \quad \tilde{I}_p = a_p \ldots a_2a_1, \quad a_k^2 = 1 \quad \text{or} \quad -1 \tag{14.15}$$

defines an isometry in E. We see the tight links between the orthogonal group $O(E)$ of the space E and its Clifford algebra $Cl(E)$.

Let κ be the number of a_k so that $(a_k)^2$ is negative. The combinations of $(-1)^p = \pm 1$ and $(-1)^\kappa = \pm 1$ give a hint on the fact that $O(E)$ is separated in 2 or 4 (following the signature) parts.

Note. There exists a proof using $Cl(E)$ (Heinz Krüger 1998, private communication), quite different of the one of Cartan-Dieudonné, of the Elie Cartan theorem by which all isometry in E is the product of symetries eachone with respect to a non isotropic hyperplan.

We have provided, on ground of the E. Cartan theorem, a very short proof of the following property: $O(E)$ may be separated into *at most* 2 or 4 *connex* components (see [3], p. 134). This proof have been completed by Henri Cartan (1984, private communication), by a proof using matricial methods in which this separation is to be considered as *at least* 2 or 4. We have then given another proof of this last property with the help of Eq. 14.15, ([3], p. 134, 135).

References

1. D. Hestenes, *Space–Time Algebra*. (Gordon and Breach, New-York, 1966)
2. R. Boudet, in *Clifford algebras and their applications in mathematical physics*, ed. by A. Micali, R. Boudet and J. Helmstetter (Kluwer, Dordrecht, 1992), p. 343
3. R. Boudet, Ann. Fond. L. de Broglie **13**, 105 (1988)

Chapter 15
Relation Between the Dirac Spinor and the Hestenes Spinor

Abstract One shows that the quaternion and the biquaternion may be considered as real forms of the Pauli and the Dirac spinors. Then relations with the real Hestenes spinor are clarified.

Keywords Pauli · Dirac · Hestenes spinor · Quaternion · Biquaternion

15.1 The Pauli Spinor and Matrices

All that follows is to be considered as relative to a frame $\{e_\mu\}$, though the vectors e_μ do not intervene directly.

1. The Pauli spinor ξ is a column ξ of two complex numbers u_1, u_2 in the form $\alpha + i\beta \in C$, where i is considered as the "imaginary number" $\sqrt{-1}$

$$\xi = \begin{pmatrix} u_1 \\ u_2 \end{pmatrix}, \quad u_1, u_2 \in C \tag{15.1}$$

2. The Pauli matrices are in the form

$$\sigma_1 = \begin{pmatrix} 0 & 1 \\ 1 & 0 \end{pmatrix}, \quad \sigma_2 = \begin{pmatrix} 0 & -i \\ i & 0 \end{pmatrix}, \quad \sigma_3 = \begin{pmatrix} 1 & 0 \\ 0 & -1 \end{pmatrix} \tag{15.2}$$

and act on the Pauli spinor ξ.

15.2 The Dirac spinor

1. The Dirac spinor is a column Ψ of four complex numbers which may be arranged into a column of two Pauli spinors

R. Boudet, *Quantum Mechanics in the Geometry of Space–Time*,
SpringerBriefs in Physics, DOI: 10.1007/978-3-642-19199-2_15,
© Roger Boudet 2011

$$\Psi = \begin{pmatrix} \xi \\ \xi' \end{pmatrix} \tag{15.3}$$

2. The Dirac matrices γ^μ are in the form

$$\gamma^0 = \begin{pmatrix} 1 & 0 \\ 0 & -1 \end{pmatrix}, \quad \gamma^k = \begin{pmatrix} 0 & \sigma_k \\ -\sigma_k & 0 \end{pmatrix} \tag{15.4}$$

and act on Ψ in such a way that

$$\gamma^0 \Psi = \begin{pmatrix} \xi \\ -\xi' \end{pmatrix}, \quad \gamma^k \Psi = \begin{pmatrix} \sigma_k \xi' \\ -\sigma_k \xi \end{pmatrix} \tag{15.5}$$

Let us consider the matrix with one line

$$\bar{\Psi} = \Psi^+ \gamma^0, \quad \Psi^+ = (\xi^+, \xi'^+) \tag{15.6}$$

where $\xi^+ = (u_1^*, u_2^*), \xi'^+ = (u_1'^*, u_2'^*)$ and u^* means the conjugate complex $\alpha - i\beta$ of $u = \alpha + i\beta \in \mathcal{C}$.

One has (see Eq. 15.15)

$$j^\mu = \bar{\Psi} \gamma^\mu \Psi \in \mathcal{R}, \quad j = j^\mu e_\mu \in M \tag{15.7}$$

In the Dirac theory of the electron whose wave function is Ψ, j is called the current of probability of presence of the electron.

Note. There exists a direct relation between the Pauli matrices and the $E^3 = \mathcal{R}^{3,0}$ space but it needs the use of the Clifford algebra $Cl(E^3)$ (see Chap. 2).

The Pauli matrices obey the same rule as the vectors e_k in $Cl(E^3)$

$$\frac{1}{2}(\sigma_i \sigma_j + \sigma_j \sigma_i) = \delta_{ij} I, \quad \mathbf{e}_i . \mathbf{e}_j = \frac{1}{2}(\mathbf{e}_i \mathbf{e}_j + \mathbf{e}_j \mathbf{e}_i) = \delta_{ij} \tag{15.8}$$

where $\{\mathbf{e}_1, \mathbf{e}_2, \mathbf{e}_3\}$ is an orthonormal frame of $\mathcal{R}^{3,0}$, the unit matrix I being identified to 1.

One deduces that there exists a direct relation between the well-known combination of the Pauli matrices

$$I, \quad \sigma_1, \sigma_2, \sigma_3, \quad \sigma_2 \sigma_3, \quad \sigma_3 \sigma_1, \quad \sigma_1 \sigma_2, \quad \sigma_1 \sigma_2 \sigma_3$$

and a frame of the Grassmann algebra $\wedge E^3$, whose dimension is $1 + 3 + 3 + 1 = 8$, composed of scalars, vectors, bivectors (or pseudo-vectors), 3-vectors (or pseudo-scalars).

In a same way the Dirac matrices obey the same rule as the vectors e_μ

$$\frac{1}{2}(\gamma_i \gamma_j + \gamma_j \gamma_i) = \delta_{ij} I, \quad e_i \cdot e_j = \frac{1}{2}(e_i e_j + e_j e_i) = \delta_{ij} \tag{15.9}$$

and there exists a direct relation between an analog combination of the Dirac matrices and a frame of $\wedge M$, whose dimension is $1 + 4 + 6 + 4 + 1 = 16$, composed of scalars, vectors, bivectors, pseudo-vectors, pseudo-scalars.

As evoked in the Preface, the identification of the matrices γ_μ to the vectors e_μ has been implicitly used by Sommerfeld [1] and Lochak [2] when they expressed the Dirac spinor by means of the Dirac matrices, but their processes corresponded implicitly to the abandon of the use of the Dirac spinor as it is defined above. That was a first approach to the clear real language of the Space–Time Algebra $Cl(M)$, introduced independently by Hestenes in [3].

It is shown in what follows (see [4]), that the Pauli and Dirac spinors are nothing else, when the matrices act upon them, but a decomposition of the Hamilton quaternion and the Clifford biquaternion, in which the number i has been replaced by the bivector of M, $e_2 \wedge e_1$ whose square in $Cl(M)$ is equal to -1.

15.3 The Quaternion as a Real Form of the Pauli spinor

Using $i = -jk$ one deduces from Eq. 2.12 the following form of a quaternion

$$q = w + kz - j(-y + kx) = u_1 - ju_2 \tag{15.10}$$

A Pauli spinor ξ, associated with the biquaternion q, is represented in the form of a column vector

$$\xi = \begin{pmatrix} u_1 \\ u_2 \end{pmatrix} \Leftrightarrow q = u_1 - ju_2 \tag{15.11}$$

that is a doublet of "complex numbers" whose the "maginary number" $\sqrt{-1}$ is nothing else but the *real* bivector $k = e_1 \wedge e_2 = e_1 e_2 = \underline{i}e_3$.

Applying (2.13) one can write, because $e_k q e_3 = -\underline{i}e_k q\underline{i}e_3$

$$e_1 q e_3 = -iqk = jk(u_1 - ju_2)k = u_2 - ju_1 \Leftrightarrow \sigma_1 \xi$$
$$e_2 q e_3 = -jqk = -j(u_1 - ju_2)k = -ku_2 - jku_1 \Leftrightarrow \sigma_2 \xi$$
$$e_3 q e_3 = -kqk = -k(u_1 - ju_2)k = u_1 - j(-u_2) \Leftrightarrow \sigma_3 \xi$$

from which one deduce the equivalences with

$$\sigma_1 = \begin{pmatrix} 0 & 1 \\ 1 & 0 \end{pmatrix}, \quad \sigma_2 = \begin{pmatrix} 0 & -k \\ k & 0 \end{pmatrix}, \quad \sigma_3 = \begin{pmatrix} 1 & 0 \\ 0 & -1 \end{pmatrix}, \quad k = \underline{i}e_3 \tag{15.12}$$

which explain the form of the σ_k matrices, and the reason why they obey the same relations as orthonormal vectors of E^3. When they act on a spinor Pauli ξ, the Hamilton quaternion q, corresponding to ξ, is to be multiplied on the left by e_1, e_2 or e_3, *and on the right* by e_3.

The conventional writing $\sqrt{-1}\xi$ is to be interpreted as

$$\sqrt{-1}\xi = \begin{pmatrix} ku_1 \\ ku_2 \end{pmatrix} = \begin{pmatrix} u_1 k \\ u_2 k \end{pmatrix}$$

and corresponds to the transformation of q into $qk = q\underline{i}e_3 = \underline{i}qe_3$.

15.4 The Biquaternion as a Real Form of the Dirac spinor

Since $\underline{i}q = q\underline{i} = q(\underline{i}e_3)e_3$

$$\Psi = \begin{pmatrix} \xi_1 \\ \xi_2' \end{pmatrix} \Leftrightarrow Q = q_1 + q_2'e_3, \quad q_2' = q_2\underline{i}e_3 \qquad (15.13)$$

that is a doublet of the Pauli spinors ξ_1, ξ_2' corresponding to q_1, q_2'.

One can write

$$e^\mu Q e_0 \Leftrightarrow \gamma^\mu \Psi \qquad (15.14)$$

that is Eq. 3.5.

Indeed, for example, since $e^0 = e_0$, $e_0 q = q e_0$, $e^0 q_2 \underline{i} e_3^2 = -q' e_3 e_0$, $e^1 = -e_1$, one has

$$e^0 Q e_0 = q_1 e_0^2 - q_2' e_3 e_0^2 = q_1 - q_2' e_3 \Leftrightarrow \gamma^0 \Psi$$

$$e^1 Q e_0 = -e_1 e^0 Q e_0 = -(e_1 q_1 e_3)e_3 + e_1 q_2' e_3 \Leftrightarrow \gamma^1 \Psi$$

One deduces in agreement with Eq. 15.7

$$\overline{\Psi}\gamma^\mu \Psi \Leftrightarrow [e^\mu \psi e_0 \tilde{\psi}]_S = e^\mu \cdot (\psi e_0 \tilde{\psi}) = e^\mu \cdot j = j^\mu \qquad (15.15)$$

that is Eq. 3.8.

References

1. A. Sommerfeld, Atombau und spectrallinien. (Fried Vieweg, Braunschweig, 1960)
2. G. Jakobi, G. Lochak, C.R. Ac. Sc. Paris **243**, 234 (1956)
3. D. Hestenes, Space–Time Algebra. (Gordon and Breach, New-York, 1966)
4. S. Gull, in *The Electron*, ed. by D. Hestenes and A. Weingartshofer (Kluwer, Dordrecht, 1991), p. 233

Chapter 16
The Movement in Space–Time of a Local Orthonormal Frame

Abstract The Dirac wave function associated with a particle contains a Lorentz rotation which allows one to deduce a local moving frame. This frame is such that some of its sub-frames play an important role in the geometrical interpretation of the gauge and the definition of a momentum-energy tensor associated with the particle. What follows is a pure geometrical description of a movement of this frame and its sub-frames independently, except the appellation of some entities, of physical considerations.

Keywords $SO^+(E)$ · Infinitesimal rotation · Local frame · Sub-frames

16.1 C.1 The Group $SO^+(E)$ and the Infinitesimal Rotations in $Cl(E)$

If both p and κ are even, denoting $I_p = R$, one can write in place of Eq. 14.15

$$y = Rx\tilde{R}, \quad \tilde{R} = R^{-1}, \quad R\tilde{R} = \tilde{R}R = 1 \tag{16.1}$$

which defines an element of the group $SO^+(E)$ whose representation in $Cl(E)$ is called $\mathrm{Spin}(E)$ (no direct relation with the term spin of quantum mechanichs).

Now consider the transform

$$b = Ra\tilde{R} \tag{16.2}$$

where $a \in E$ is constant and R is a function of a variable t. We can write

$$\frac{db}{dt} = \frac{dR}{dt}a\tilde{R} + Ra\frac{d\tilde{R}}{dt} = \frac{dR}{dt}\tilde{R}Ra\tilde{R} + Ra\tilde{R}R\frac{d\tilde{R}}{dt}$$

R. Boudet, *Quantum Mechanics in the Geometry of Space–Time*,
SpringerBriefs in Physics, DOI: 10.1007/978-3-642-19199-2_16,
© Roger Boudet 2011

Denoting

$$\Omega = 2\frac{dR}{dt}\tilde{R} = -2R\frac{d\tilde{R}}{dt} \qquad (16.3)$$

we can write

$$\frac{db}{dt} = \frac{1}{2}(\Omega b - b\Omega)$$

$$\frac{db}{dt} = \Omega \cdot b, \quad \Omega \in \wedge^2 E \qquad (16.4)$$

This relation and the fact that Ω is a bivector are immediately deduced from Eqs. 14.7, 14.1. Ω is called the bivector which defines the infinitesimal rotation associated with the rotation Eq. 16.2.

Equation 2.2 gives $(\Omega \cdot b) \cdot b = \Omega \cdot (b \wedge b) = 0$ and confirms the orthogonality of $\Omega.b$ and db/dt deduced from the derivation of the constant b^2.

16.2 Study on Properties of Local Moving Frames

Similar studies may be made for all euclidean space. Here we are more particurlarly interested in $M = R^{1,3}$, but, for the while, all that follows in this section is quite independent, except the appellation of some entities, of physical considerations.

16.3 Infinitesimal Rotation of a Local Frame

We consider a fixed positive orthonormal frame $\{e_\mu\}$ of M so $e_0^2 = 1$ (timelike) and $e_k^2 = -1$ (spacelike), $k = 1, 2, 3$, or galilean frame, and denote $x = x^\mu e_\mu = x_\mu e^\mu$, $(e^\nu e_\mu = \delta_\mu^\nu)$ the current point of M.

Let us consider $R \in \mathrm{Spin}(M)$ (Lorentz rotation) depending on $x = x^\mu e_\mu \in M$ with at least double derivatives with respect to $\partial_\mu = \partial/\partial x^\mu$.

We will consider the local orthonormal frame

$$F = \{v, n_1, n_2, n_3\}, \quad v = Re_0\tilde{R}, \quad n_k = Re_k\tilde{R} \qquad (16.5)$$

or Takabayasi–Hestenes frame associated with a particle.

The infinitesimal rotation of this frame, associated with the variation of the point x, is defined by the bivectors

$$\Omega_\mu = 2\partial_\mu R\tilde{R} \qquad (16.6)$$

which satisfies the property, deduced from $\partial_{\mu\nu}^2 R = \partial_{\nu\mu}^2 R$

$$\partial_\nu \Omega_\mu - \partial_\mu \Omega_\nu + \frac{1}{2}(\Omega_\mu \Omega_\nu - \Omega_\nu \Omega_\mu) = 0 \tag{16.7}$$

16.4 Infinitesimal Rotation of Local Sub-Frames

We are interested in the infinitesimal rotations of sub-frames of F upon themselves in the motion of F associated with the variation of the point x.

1. The infinitesimal rotation of the oriented plane $\{n_2, n_1\}$ (which is considered in the state "up" of the electron, $\{n_1, n_2\}$ being considered in the state "down") may be defined (see [1]) by the vector, using Eq. 16.4,

$$\omega = \omega_\mu e^\mu, \quad \omega_\mu = \Omega_\mu \cdot (n_2 \wedge n_1) = \partial_\mu n_2 \cdot n_1 = -\partial_\mu n_1 \cdot n_2 \tag{16.8}$$

This infinitesimal rotation appears in the infinitesimal rotation of the sub-frame $\{v, n_2, n_1\}$ upon itself in "up", or $\{v, n_1, n_2\}$ in "down".

We have introduced in [2] the linear application N of M into M (tensor)

$$n \in M \to N(n) = (\Omega_\mu \cdot (\underline{i}(n_3 \wedge n)))e^\mu \in M \tag{16.9}$$

Because of the relations $\underline{i} = v n_1 n_2 n_3$ and

$$\underline{i}(n_3 \wedge v) = n_2 \wedge n_1, \quad \underline{i}(n_3 \wedge n_1) = n_2 \wedge v, \quad \underline{i}(n_3 \wedge n_2) = v \wedge n_1$$

$$\Omega_\mu \cdot (l \wedge m) = (\Omega_\mu \cdot l) \cdot m = (\partial_\mu l) \cdot m$$

we can write $N(n_3) = 0$ and

$$N(v) = (\partial_\mu n_2 . n_1)e^\mu, \quad N(n_1) = (\partial_\mu n_2 . v)e^\mu, \quad N(n_2) = (\partial_\mu v . n_1)e^\mu \tag{16.10}$$

The tensor N expresses the infinitesimal rotation of the sub-frame $\{v, n_2, n_1\}$ upon itself.

This tensor appears in the definition of the momentum–energy tensor of the electron in a form such that

$$\frac{1}{2}N(n) = e^\mu[(\partial_\mu R)\underline{i}e_3\tilde{R}]_S \tag{16.11}$$

2. In the same way we have defined in [3] the tensor

$$S(n) = (\Omega_\mu . (\underline{i}(v \wedge n)))e^\mu \tag{16.12}$$

and so $S(v) = 0$ and

$$S(n_1) = (\partial_\mu n_2 \cdot n_3)e^\mu, \quad S(n_2) = (\partial_\mu n_3 \cdot n_1)e^\mu, \quad S(n_3) = (\partial_\mu n_1 \cdot n_2)e^\mu \quad (16.13)$$

The tensor S expresses the infinitesimal rotation of the sub-frame $\{n_1, n_2, n_3\}$ upon itself.

This tensor appears in the definition of the momentum-energy tensors in theories using the $SU(2)$ gauge in a form such that

$$S(n) = \frac{1}{2}e^\mu [\partial_\mu R\underline{i}e_0\tilde{R}]_S \tag{16.14}$$

16.5 Effect of a Local Finite Rotation of a Local Sub-Frame

For reasons which appear in the physical theories and which will be alluded to below, we are going to complicate the above considerations by supposing that the rotation R is change into RU where U is a finite rotation upon itself of a sub-frame of F. The group of the rotation U is called a gauge, a local or a global gauge following the cases where the rotations U are or not considered as depending on the point x.

(1) In what is called the U(1) gauge in the complex language a change of gauge corresponds in STA to

$$U = \exp(e_2 e_1 \chi/2), \quad R \to R' = RU = R\exp(e_2 e_1 \chi/2) \tag{16.15}$$

which induces a rotation through an angle χ in the plane (n_2, n_1):

$$n_2' = \cos\chi\, n_2 + \sin\chi\, n_1, \quad n_1' = -\sin\chi\, n_2 + \cos\chi\, n_1 \tag{16.16}$$

with $R'e_0 R'^{-1} = e_0$, $R'n_3 R'^{-1} = n_3$.

(2) In the SU(2) gauge, a change of gauge corresponds in STA to

$$U \in \mathrm{Spin}(M) : Ue_0U^{-1} = e_0, \quad R \to R' = RU \Rightarrow R'e_0 R'^{-1} = v \tag{16.17}$$

giving

$$\hat{\Omega}_\mu = 2(\partial_\mu U)U^{-1}, \quad \Omega_\mu \to \Omega_\mu' = \Omega_\mu + R\hat{\Omega}_\mu R^{-1} \tag{16.18}$$

The change leaves v invariant but defines a rotation of the sub-frame upon $\{n_1, n_2, n_3\}$ upon itself.

Applying Eq. 16.7 to U, one deduces that the $\hat{\Omega}_\mu$ verify, if the gauge is local

$$\partial_\nu \hat{\Omega}_\mu - \partial_\mu \hat{\Omega}_\nu + \frac{1}{2}(\hat{\Omega}_\mu \hat{\Omega}_\nu - \hat{\Omega}_\nu \hat{\Omega}_\mu) = 0 \tag{16.19}$$

References

1. R. Boudet , C.R. Ac. Sc. (Paris) **272** A, 767 (1971)
2. R. Boudet , C.R. Ac. Sc. (Paris) **278** A, 1063 (1974)
3. R. Boudet, in *Adv. appl. Clifford alg.* (Birkhaüser Verlag, Basel, Switzerland, 2008), p. 43

Chapter 17
Incompatibilities in the Use of the Isospin Matrices

Abstract Examples are given of the non possibility of the use of isospin matrices when they act on a Dirac spinor or a couple of Dirac spinors. But they can act on a right or a left doublet. In this case a faithful translation in the real language is possible.

Keywords Dirac spinor · Doublet · Isospin matrices

17.1 Ψ is an "Ordinary" Dirac Spinor

1. Consider for example each numbers $\bar{\Psi}\gamma^0\tau_k\Psi$

For $k = 3$, the number is real but for $k = 1, 2$ the numbers are purely imaginary and cannot be associated with the numbers W_0^1, W_0^2 which are real.

17.2 Ψ is a Couple (Ψ_a, Ψ_b) of Dirac Spinors

If each Ψ_a, Ψ_b are "ordinary" Dirac spinors (that is corresponding to invertible biquaternions), the strict application of the complex formalism associates with each real vector W_μ^k in the form

$$j_{ab}^\mu = \bar{\Psi}_a\gamma^\mu\Psi_b + \bar{\Psi}_b\gamma^\mu\Psi_a, \ k_{ab}^\mu = i(\bar{\Psi}_b\gamma^\mu\Psi_a - \bar{\Psi}_a\gamma^\mu\Psi_b)$$

$$j_a^\mu - j_b^\mu = \bar{\Psi}_a\gamma^\mu\Psi_a - \bar{\Psi}_b\gamma^\mu\Psi_b, \tag{17.1}$$

respectively.

The association of these vectors with the W_μ^k seems incompatible with the role of the τ_k as they have been presented in the Y. M. theory of the $SU(2)$ gauge and so the form of the bivector field associated with vector bosons.

R. Boudet, *Quantum Mechanics in the Geometry of Space–Time*,
SpringerBriefs in Physics, DOI: 10.1007/978-3-642-19199-2_17,
© Roger Boudet 2011

17.3 Ψ is a Right or a Left Doublet

The τ_k are considered as acting upon components in the form

$$\Psi_l^\pm = \frac{1}{2}(1 \pm \gamma^5)\Psi_l, \quad \gamma^5 = \gamma^0\gamma^1\gamma^2\gamma^3 i, \quad l = 1, 2 \tag{17.2}$$

where Ψ_l are "ordinary" Dirac spinors, that is which correspond to invertible bi-quaternions.

In this case the τ_k may be considered as matrices

These components correspond in STA to *non invertible* biquaternions $\psi_l^\pm$, deduced from invertible biquaternions ψ_l, which are arranged in a biquaternion ψ which may be invertible (Eq. 9.13):

$$\psi_l^\pm = \psi_l \frac{1}{2}(1 \pm \mathbf{e}_3), \quad \psi = \psi_1^\pm \pm \psi_2^\pm \mathbf{e}_1 \tag{17.3}$$

We emphasize that in this case our transposition is a step by step translation in STA of the complex formalim.

17.4 Questions about the Nature of the Wave Function

We recall what we have established concerning the Y. M. theory.

There is no problem in the fact that the wave function Ψ, on which the τ_k matrices act, is an "ordinary" Dirac spinor, but not if it is a couple of Pauli or Dirac spinors (as, in this last case, in the chromodynamics theory). As far as that a left or a right doublet Ψ imposes to this spinor to be invertible, it seems that the only possibility about the nature of Ψ is to be such a doublet.

So, in the electroweak theory, confirmed by the experiment, where Ψ is a left doublet, the question is solved. But it remains in all the theories, such as the chromodynamics one, in which the τ_k matrices are used.

Chapter 18
A Proof of the Tetrode Theorem

Abstract The STA Krüger proof is much more shorter than the one of Tetrode. (That does not take out the merit of this proof achieved just after the publication of the Dirac equation!)

Keywords Hestenes spinor · Tetrode tensor

What follows is a STA proof due to Heinz Krüger (private communication, 2010). To be in agreement with the general method of presentation of the theorems we use in the present book, our writing of the proof is a bit longer than the Krüger one.

One uses the Dirac equation in the form

$$\hbar c e^{\mu}(\partial_{\mu}\psi)\underline{i}e_3\tilde{\psi} = (mc^2\psi + qA\psi e_0)\tilde{\psi} \tag{18.1}$$

One chooses the vectors n as vectors e^{ν} of the laboratory frame

$$T(e^{\nu}) = \hbar c e^{\mu}\left[e^{\nu}(\partial_{\mu}\psi, \underline{i}e_3\tilde{\psi}\right]_S - (e^{\nu} \cdot j)A \in M \tag{18.2}$$

where $j = q\rho v$ and one writes

$$T^{\nu} = T(e^{\nu}) = e^{\mu}I - (e^{\nu} \cdot j)A, \quad I = \hbar c\left[e^{\nu}(\partial_{\mu}\psi)\underline{i}e_3\tilde{\psi}\right]_S \tag{18.3}$$

One deduces

$$\partial_{\nu}T^{\nu} = e^{\mu}\partial_{\nu}I - (e^{\nu} \cdot j)\partial_{\nu}A - (e^{\nu} \cdot \partial_{\nu}j)A = e^{\mu}\partial_{\nu}I - (e^{\nu} \cdot j)\partial_{\nu}A \tag{18.4}$$

since the conservation of the current implies $e^{\nu} \cdot \partial_{\nu}j = 0$.
One can write

$$e^{\mu}\partial_{\nu}I = (\partial_{\nu}J + \partial_{\nu}K)e^{\mu} \tag{18.5}$$

with

$$\partial_\nu J = \hbar c \left[e^\nu (\partial_\nu(\partial_\mu \psi)) \underline{i} e_3 \tilde{\psi} \right]_S = \hbar c \left[\partial_\mu (e^\nu (\partial_\nu \psi) \underline{i} e_3) \tilde{\psi} \right]_S \qquad (18.6)$$

since $\partial_\nu(\partial_\mu X) = \partial_\mu(\partial_\nu X)$, and applying Eq. 18.1 with the notation $e^\nu \partial_\nu$ in place of $e^\mu \partial_\mu$

$$\partial_\nu J = \left[(\partial_\mu (mc^2 \psi + q A \psi e_0)) \tilde{\psi} \right]_S$$

$$\partial_\nu J = \left[(mc^2 (\partial_\mu \psi) \tilde{\psi} \right]_S + \left[q A (\partial_\mu \psi) e_0 \tilde{\psi} \right]_S + (\partial_\mu A) \cdot j \qquad (18.7)$$

and

$$\partial_\nu K = \hbar c \left[e^\nu (\partial_\mu \psi) \underline{i} e_3 (\partial_\nu \tilde{\psi}) \right]_S = \hbar c \left[(\partial_\mu \psi) \underline{i} e_3 (\partial_\nu \tilde{\psi}) e^\nu \right]_S$$

$$\partial_\nu K = \hbar c \left[e^\nu (\partial_\nu \psi) e_3 \underline{i} (\partial_\mu \tilde{\psi}) \right]_S \qquad (18.8)$$

Applying again Eq. 18.1 but with $e_3 \underline{i} = -\underline{i} e_3$, one has

$$\partial_\nu K = \left[(-(mc^2 \psi - q A \psi e_0)(\partial_\mu \tilde{\psi}) \right]_S$$

$$\partial_\nu K = - \left[mc^2 (\partial_\mu \psi) \tilde{\psi} \right]_S - \left[q A (\partial_\mu \tilde{\psi}) e_0 \psi \right]_S$$

with

$$- \left[q A (\partial_\mu \tilde{\psi}) e_0 \psi \right]_S = - \left[q (\partial_\mu \tilde{\psi}) e_0 \psi A \right]_S = - \left[q A (\partial_\mu \psi) e_0 \tilde{\psi} \right]_S \qquad (18.9)$$

where $[aX]_S = [Xa]_S, a \in M, [X]_S = [\tilde{X}]_S$, have been applied for the writing of Eqs. 18.8 and 18.9.

One can write $((\partial_\mu A) \cdot j)e^\mu = ((\partial_\nu A) \cdot j)e^\nu$ and using Eq. 2.1 one has

$$\partial_\nu T^\nu = ((\partial_\nu A) \cdot j)e^\nu - (e^\nu \cdot j)\partial_\nu A = (e^\nu \wedge \partial_\nu A) \cdot j = F \cdot j \qquad (18.10)$$

and at least

$$\partial_\nu T^\nu = \rho(q F.v) \qquad (18.11)$$

Chapter 19
About the Quantum Fields Theory

Abstract In QFT the potentials are put in a complex form with the association $\hbar i$ of $\hbar$ with the number $i = \sqrt{-1}$ by analogy with the presence of $\hbar i$ in the Dirac equation. But in this equation i has in fact the meaning of a real bivector of space-time which has no place in an electromagnetic potential. The results are the same as a real quantum electrodynamics because, in the calculations, the imaginary parts of these potentials are null. But when the potential is in the form q/R, the QFT construction leads to the presence of an unacceptable nonsense.

Keywords Complex potentials · Plank constant · Number i

19.1 On the Construction of the QFT

The quantum fields theory (QFT) was built on the basis of the works of Dirac, Jordan and Pauli, Heisenberg and Pauli, during the years 1927–1929. It seems that its purpose is to express as closely as possible the photon like a particle.

Here we are only interested in its application by the points of the theory which have been used in the Lamb shift calculation.

For describing the two ways of the construction of the QFT leading to this calculation, we follow the treatise of Heitler [1], considered, at least for a long time, as the usual reference for the statement of the QFT.

1. A Mathematical Construction

One considers a "*vector potential* **A** *which ... may be written as a series of plane waves*" ([1] Para 7, p. 56, lines 1–3)

$$\mathbf{A} = \mathbf{A}_0 + \mathbf{A}_0^* \tag{19.1}$$

where $\mathbf{A}_0$ is complex and $\mathbf{A}_0^*$ is the complex conjugate of $\mathbf{A}_0$ ([1], Para 6, Eq. 14). In such a way that **A** is *real* in agreement with "the classical radiation theory" ([1],

R. Boudet, *Quantum Mechanics in the Geometry of Space–Time*,
SpringerBriefs in Physics, DOI: 10.1007/978-3-642-19199-2_19,
© Roger Boudet 2011

Para 6) and so the laws of the classical electromagnetism may be respected, at least for a potential which may be written as a series of plane waves.

2. A rule presented as physical

The imaginary number $i = \sqrt{-1}$ is replaced, in the application of the previous construction to some operators, by $i\hbar$ where $\hbar$ is the reduced Plank constant. This replacement is justified *"by exact analogy with the ordinary quantum theory"* ([1], Para 7, p. 56, lines 10–11).

The origin of the above rule lies in the fact that $\hbar$ appears in the electromagnetic fields in quantum theory, and that, in the quantum theory of the electron, the number i appears in the association $i\hbar$, with $\hbar$.

Though the replacement is placed in operators ([1], Para 7, Eq. 6), as a necessity, it is translated in the expression of the potentials given by Eq. 19.1 which is is considered as "quantified".

It is exactly in this way, association $i\hbar$ of $\hbar$ with i, that the potentials appear in the standard Lamb shift calculation.

19.2 Questions

1. The association of $\hbar$ with i.

This association, "by exact analogy with the ordinary quantum theory" calls we the following question. The product $i\hbar$ appears in the Dirac equation of the electron. But in this equation the meaning of "i" is not the imaginary number $\sqrt{-1}$ but a bivector of Space–Time $e_2 \wedge e_1 = e_2 e_1$ (or $e_1 e_2$, following the two possible orientations of the spin), whose square in STA is equal to -1 and so a real object.

We recall that this explanation of Hestenes in [2] was in some way already included in the works of Sommerfeld [3] and Lochak [4] in which i was replaced by $\gamma_2\gamma_1$ the Dirac matrices γ_μ being implicitly identified to the vectors e_μ of a galilean frame.

The presence of a bivector in a electromagnetic potential cannot be considered. So the "exact analogy with the ordinary quantum theory", that is mostly at this time the Dirac equation of the electron, seems not appropriate.

2. The decomposition of a potential q/r in plane waves.

Such a decomposition is the source of an artifice, unseen as well by the physicists who have used it as the ones who use the QFT. We have pointed out this artifice in [5].

The use of this artifice does not alter the results in the calculation of the Lamb shift. But it shows that the QFT, despite its mathematical correctness, cannot be be considered as a physical theory when it is applied to the Lamb shift, though this calculation is still considered as an "outstanding triumph of the QFT".

19.3 An Artifice in the Lamb Shift Calculation

The formula in Heitler [1], Para 34, Eq. (4 $'$), which allows the calculation of one of the three terms, the *Electrodynamics static term* W_S, which compose the shift, is the following:

$$\frac{\left(\Psi_0^*(\mathbf{r})\Psi_n(\mathbf{r})\right)\left(\Psi_n^*(\mathbf{r}')\Psi_0(\mathbf{r}')\right)}{|\mathbf{r}-r'|}$$
$$= \frac{1}{2\pi^2\hbar c}\int[(\Psi_0^*(\mathbf{r})e^{i(\mathbf{k}.\mathbf{r})/\hbar c}\Psi_n(\mathbf{r}))(\Psi_n^*(\mathbf{r}')e^{-i(\mathbf{k}.\mathbf{r}')/\hbar c}\Psi_0(\mathbf{r}'))]\frac{d^3k}{k^2} \quad (19.2)$$

Multiplied by $e^2/2$ (see [1], Para 34, Eq. 43) this formula expresses the contribution W_S of an electron in a state of energy E_n to the shift of the same electron in a state of energy E_0.

It implies a static potential e/R, where $R = |\mathbf{r} - r'|$ corresponds to the spatial positions associated with these two states.

One can observe that $\hbar c$ is not in the left hand part of the formula but is present in the right one. And one deduces that the following equality has been used

$$\frac{1}{|\mathbf{r}-\mathbf{r}'|} = \frac{1}{2\pi^2\hbar c}\int e^{i((\mathbf{k}.(\mathbf{r}-r'))/\hbar c)}\frac{d^3k}{k^2} \quad (19.3)$$

where

$$d^3k/k^2 = \sin\theta d\theta d\varphi dk = d\Omega dk$$

and where the presence of a factor $1/\hbar c$ in Eq. 19.2 is due to the fact that k has the dimension of an energy because in $\exp[i((\mathbf{k}/\hbar c) \cdot (\mathbf{r} - r'))]$ the vector $\mathbf{k}/\hbar c$ must have the dimension the inverse of a length, in such a way that, after the integration, the dimension of the right part of Eq. 19.3 is the inverse of a length as its left part.

Equation 19.3 is in full agreement with the construction of the QFT: decomposition of the potential in plane waves (?) by the use of the so called "Fourier transform" of $1/R$ (see [6], Eq. 16)

$$\frac{1}{|\mathbf{r}-\mathbf{r}'|} = \frac{1}{2\pi^2}\int e^{i(\mathbf{k}.(\mathbf{r}-r'))}\frac{d^3k}{k^2} \quad (19.4)$$

apparent complex form of the potential, and with the transform $i/\hbar = -1/i\hbar$ (allowed by the fact that the imaginery part of (19.3) cancels), association $i\hbar$ of i with $\hbar$.

Note that in the article of Kroll and Lamb [6], Eq. 19.4 is identical to Eq. 19.3 from the fact that one writes in this article $\hbar = c = 1$ (p. 392) and so this notation is applied to the Eq. 23, which follows this writing of $\hbar$ and c, giving W_S (see also the articles of Dyson, French and Weisskopf of 1949).

We recall the calculation that we have made in [5] in which we have pointed out the artifice.

The only explanation of the presence of $\hbar c$ in the right hand part of Eq. 19.3 is the following.

This presence corresponds to the construction.

$$\frac{1}{R} = \left(\frac{\pi}{2} \times \frac{1}{R}\right) \times \frac{2}{\pi} \tag{19.5}$$

then

$$\frac{\pi}{2} = \int\limits_0^\infty \frac{\sin x}{x} dx = \int\limits_0^\infty \frac{\sin(k'R)}{k'} dk'$$

Denoting $k' = k/\hbar c$

$$\frac{\pi}{2} = \int\limits_0^\infty \frac{\sin(kR/\hbar c)}{k'} dk' \tag{19.6}$$

and using

$$\frac{\sin(k'R)}{k'R} = \frac{1}{2} \int\limits_0^\pi e^{ik'R\cos\theta} \sin\theta d\theta = \frac{1}{4\pi} \int\limits_0^\pi \int\limits_0^{2\pi} e^{i(\mathbf{k'.R})} d\Omega$$

one deduces

$$\frac{1}{R} = \left[\frac{1}{4\pi} \int\limits_0^\infty \int\limits_0^\pi \int\limits_0^{2\pi} e^{i(\mathbf{k'.R})} dk' d\Omega \right] \times \frac{2}{\pi} = \frac{1}{2\pi^2 \hbar c} \int e^{i((\mathbf{k.R})/\hbar c)} \frac{d^3k}{k^2}$$

that is the formula Eq. 19.3.

One sees on Eqs. 19.5, 19.6 *that the Planck constant is introduced inside $\pi/2$ (not inside $2/\pi$), an indisputable nonsense!*

Nevertheless the use of such a device does not alter the validity of the calculation of the Lamb shift which remains one of the most admirable work in the theory of the electron.

References

1. W. Heitler, *The Quantum Theory of Radiation* (Clarendon Press, Oxford, 1964)
2. D. Hestenes, J. Math. Phys. **8**, 798 (1967)
3. J. Seke, Mod. Phys. Lett. B **7**, 1287 (1993)
4. G. Jakobi, G. Lochak, C.R. Ac. Sc. (Paris) **243**, 234 (1956)
5. R. Boudet, *New Frontiers in Quantum Electrodynamics and Quantum*, ed. by A. O. Barut (Plenum, New York, 1990), p. 443
6. M. Kroll, W. Lamb, Phys. Rev. **75**, 388 (1949)

Index